EDWARD CLODD

PRIMITIVE MAN

Elibron Classics
www.elibron.com

Elibron Classics series.

ISBN 0-543-96133-8 (paperback)
ISBN 0-543-96132-X (hardcover)

This Elibron Classics Replica Edition is an unabridged facsimile of the edition published in 1909 by George Newnes, Ltd., London.

FIG. 1.

Copyright 1894, *Gabr. Max, pinx.*] [*Hanfstaengl, Munich.*

ANCESTORS OF MAN (*Pithecanthropus Alalus*).

PRIMITIVE MAN

BY

EDWARD CLODD

WITH ILLUSTRATIONS

GEORGE NEWNES, LTD.

LONDON MCMIX.

PREFATORY NOTE.

The List of Books given at the end of this little volume fulfils the twofold purpose of indicating the authorities who have been consulted in its preparation, and of telling the reader where fuller information on the several subjects dealt with is to be found.

My special acknowledgments are due both to Sir John Evans and Messrs. Longmans and Co. for their generous permission of the use of blocks from Sir John Evans's "Ancient Stone Implements of Great Britain," and "Ancient Bronze Implements of Great Britain."

E. C.

19, Carleton Road,
Tufnell Park, N.
February, 1895.

Dedicated (BY PERMISSION)

TO

SIR JOHN EVANS, K.C.B., D.C.L., LL.D.,

V.P.R.S., ETC., ETC.,

WHOSE LABOURS IN THE COLLECTION
OF EVIDENCE BEARING ON THE
ANTIQUITY AND PRIMITIVE STATE OF MAN
HAVE PLACED ALL
STUDENTS OF ANTHROPOLOGY UNDER
LASTING INDEBTEDNESS.

CONTENTS.

CHAPTER I.

THE PLACE OF MAN IN THE EARTH'S LIFE-HISTORY.

In no branch of knowledge has there been more rapid advance during the past fifty years than in that which deals with the history of so-called "primitive" man. A generation or two ago, inquiry was rarely pushed beyond the sources of information supplied by written documents, coins, inscriptions, and such like materials. The possible existence of other materials throwing light on remote ages in which man had played a part—ages about which history was either silent, or recorded only myths and legends—was but here and there recognised.

Apropos of "antiquarian research," Boswell reports Johnson as saying: "All that is really *known* of the ancient state of Britain is contained in a few pages. We *can* know no more than what the old writers have told us." That remark gauges the high-water level reached a little more than a century ago. And yet the evidence which Dr. Johnson declared non-existent was beneath the soil of his beloved London. Even at the time he spoke, there was lying in the Sloane Museum a rudely-chipped flint weapon which had been found, at the end of the seventeenth century, associated with an elephant's tooth "opposite to black Mary's near Grayes inn lane," in which street once lived Johnson. But years passed before it was known that these and

FIG. 2.—FLINT IMPLEMENT FROM GRAY'S INN LANE. (*Evans.*)

other unheeded relics epitomised the early history of man and the condition of the Thames valley when he and a strange group of animals lived there in a dim and dateless past.

Until a few years ago our school histories of Britain began with the invasion of Julius Cæsar. Both these and graver histories were silent about man and his doings before that time. It seemed not to occur to the writers to inquire whether the "ancient Britons" were the original inhabitants of these islands; and if not, whence they came, and who lived here before them; and if the earlier folk had left any remains from which something could be known as to their civilised or barbaric state. All this is changed, and for the better. A new science has been born—the science of man. In other words, the same method of inquiry which is applied to origins generally is applied to him. He stands as an exception no longer: he is included in the universal order.

The name "pre-historic" has been given to the vast period about which written records tell us nothing, because it lies outside the horizon of history, as we define it. There have been unearthed from ancient river-beds, limestone caverns, lake-bottoms, and refuse-heaps; from rude sepulchres and stone structures an enormous mass of relics which reveal to us the story of man during periods when the Continent of Europe stretched beyond Great Britain and Ireland into the Atlantic, and was joined at more than one point to Africa.

Besides the knowledge gained from these relics of man's presence, much has been gathered in recent years about the blood-relationship of various races, and, more than this, about man's place in the long chain of life on the globe.

Therefore, not only has inquiry into his history been carried back to periods not to be reckoned by years, but he is no longer treated as a being apart from other living things. And that is what is meant by the science of man which we call "anthropology."

A few elemental facts will help to make clear his place in the order and succession of life.

The material basis and vehicle of life is a slimy stuff called "protoplasm," which builds for itself a thin-walled cell. Every plant and animal is made up of cells, the shape and arrangement of which are governed by the work which they have to do. The lowest life-forms consist of one cell, which does everything; that is, takes-in food and air, casts out refuse, and responds to its surroundings. All life-forms above the lowest consist of many cells, of which the several parts of the body are built up, each part doing its own work on the principle of division of labour. The one-celled multiply by division; the lower many-celled by the congregation of like cells; the higher by the more complex fusion of unlike cells, as the sperm-cell of the male with the germ-cell of the female, giving rise to a seed or egg whence grows offspring resembling the parents.

The Vertebrates, or back-boned animals, stand at the head of the many-celled; the Mammals, or those that suckle their young, at the head of the Vertebrates; and the order of the Primates (pronounced Pri-ma-tes, to distinguish them from archbishops) at the head of the Mammals. This order includes lemurs. monkeys, apes, and man.

The anthropoid or man-like apes—the gibbon, orang-outang, chimpanzee, and gorilla—are man's nearest allies. Some of them resemble him more in one feature; some in another. The orang-outang has the most human-like brain; the chimpanzee has the most human-like skull; and the more savage gorilla has the most human-like feet and hands. Although the bones of a man cannot be mistaken for those of an anthropoid ape, the skeleton of each, bone for bone, are identical. If we compare the skull of a horse with a human skull, we find the same number of bones. And whether it be man, or ape, or horse, depends not on differences in the plan of the general skeleton, but in the proportions, as, for example, dealing only with the skulls of each, in the size of the brain-case and the face. For the comparisons of structure make clear that all differences are of degree, not of kind. The lower apes vary more, especially in their brains, from the highest apes than these differ from man.

The barest summary of the evidence in proof of the descent of all living things from a common ancestry is not possible, neither needful, here, and it must therefore suffice to say that the common descent of man and apes is no longer to be doubted. But man is neither the offspring nor the brother of the apes; he is a sort of cousin more than "once removed." And the answer to the oft-put question, Where is the missing link between them? is, There is no missing link; there never has been one. As with the likenesses and differences between the apes themselves, so with those between apes and man. The likenesses are explained by descent from a common ancestry; the differences have slowly arisen in subtle ways. The Primates form the upper branches of the life-tree, whose highest branch is man. This topmost place has been won by him in virtue of certain advantages in his bodily structure, namely, his wholly erect posture, his hands, and his organs of speech. For, although the impassable gulf between man and apes is especially manifest in his larger and more deeply furrowed brain, this is more an effect than a cause of the advantages just named. That without them there would not have been developed such difference of brain a brief explanation will show.

The primary use of limbs is to enable an animal to move about in search of food This is accomplished in various ways among the lower life-forms; and, among the Vertebrates, by two pairs of limbs. These have been modified in various animals for different modes of action, as in the fore-limbs of the bat and bird for flight; in the flipper of the whale for propelling it through the water; and so on. But in these cases, the fore-limbs remain organs of locomotion. And they remain so among quadrupeds and the man-like apes. It is true that the gibbon can walk erect, but his gait is waddling and inconstant, his habits being arboreal; and his long arms used in his wide leaps from tree to tree. The other big apes are only semi-erect. Man alone has acquired the wholly upright position which has set his fore-limbs entirely free to act as organs for handling, grasping, and throwing things. The modification of the fingers, enabling them to be opposed singly, or all together, to the thumb, and thereby to act as hooks or clasps; to form a cuplike palm; to grasp things large or small, and thus learn something about them; gave man a perfect organ without which he could never have won lordship over the earth. And we have but to cripple or lose a thumb to realise that in it lies the real power of the hand. The prehensile or grasping organs of some of the lower animals, as of the elephant, monkey, parrot, and opossum, whereby they can lay hold of an object and learn something of its nature, raises them in the scale of intelligence; and when we contrast trunk or claw with the human hand, we see what a mighty agent this has been in development of brain. Obviously the attainment of the erect posture involved various changes in the structure of man's body—as the thickening of the leg bones, the flattening of the feet, the curve of the spine, and the altered position of the skull as balanced on it. In all this there was the making of Man. His two-footed and upright posture involved exchange of the tree-life of his ancestors for life on the ground, which brought him into new relations with his surroundings, and finally, in the ceaseless struggle for life which he had to wage, gave him the mastery over foes and the wide earth itself. That the wholly-erect posture was acquired late, speaking relatively, in man's development from an ape-like ancestry is shown, among other ways, in the crawling of infants for some time after birth—which shows the quadrupedal instinct—and in the preference we all have for sitting down. Among civilised people the great toe is, not infrequently, opposable, like the great toe in apes; the Chinese can row with it, and the lower races use it for grasping. It has been ingeniously pointed out that one of the many proofs of man's descent from a tree-dwelling ancestry is in his behaviour when he is in danger of drowning. He acts in the water as if trying to scramble to a place of safety, extending his arms above him as in climbing.

Man's acquisition of articulate speech is in itself sufficing evidence of his social habits. For language is wholly and strictly a social institution; man speaks to impart his thoughts; a solitary man would not have developed a language, since the need for it would not have arisen. In the degree that animals are gregarious, they are higher in the scale—as ants, bees, and wasps among insects; dogs, elephants, &c., among mammals; and the instincts which led the apes and man to their several social ways of life were inherited from their common ancestors, and strengthened by practice, being, in fact, necessary to their existence, and to the successful rearing of their offspring.

The normal state of every living thing, from the lowest plant to the highest animal, is one of conflict, and the "weakest go to the wall." Unlike the lion and other beasts of great muscular strength, the Primates had no powerful organs of attack or defence, and so took to living in trees, where their grasping organs stood them in good stead. Swiftness of motion was their safety; the need for alertness against wilder and fiercer beasts not only quickened their wits, but compelled them to unity. Whichever among them possessed any favourable variation, no matter how slight, in structure of brain or sense-organs or pliability of fore-limb, would secure an advantage over less-favoured rivals in the common struggle for life, and, transmitting their advantages to their offspring, would in the long run wholly distance their competitors. Thus may be explained the advance of man's progenitors over those of the highest apes. But, in the case of both man and ape, social bonds were strengthened by the dependence of their offspring. Among the lower animals the offspring are hatched or born fully equipped, so that their parents need trouble little about them. But among the highest animals the offspring, for a longer or shorter period after birth, are helpless, and dependent on the parents. That condition evokes the nurturing care of father and mother, as well as the sympathy and love which helplessness excites; hence the development of social relations, which, beginning with the family, extend to groups of families out of which are formed tribes and nations. The longer the infant stage, the more intelligent the animal, which explains why babyhood is longer in the offspring of man than of apes, and of apes than of monkeys.

We may now, perhaps, better understand the causes which impel the development of articulate speech—indeed, of inarticulate speech also. For the social animals communicate with one another by sounds, which convey certain meanings. A dog can often understand what we say to him, and therefore it is more than probable that he can still better understand what a fellow-dog means by a certain bark. As between the language of man and the language of animals the difference is one of degree; there is no mystery in either: the faculty of speech lies in brain and larynx, these organs undergoing marked modification in man at a very early period in his history.

Concerning the beginnings of speech, the Roman poet Lucretius made a shrewd guess when he said, "Nature impelled men to utter various sounds of the tongue, and use struck out the names of things." For, when we pull words to pieces, a very large number are found to have their origin in sounds which are imitations of natural sounds or instinctive cries, whose currency depended on their success in conveying the meaning intended by their users. No mystic bond linked word and thought together; utility and convenience alone joined them. Of course, grimaces and gestures like those still largely employed by both barbaric and demonstrative races, played no small supplemental part at the outset.

While the language of the lower animals remains at the instinctive stage and is untransmitted; human speech, in addition to transmission, preserves, through the art of writing, the wisdom and experience of one generation for the benefit of the next, so that each starts where its forefathers left off, and in turn adds to the garnered intellectual wealth of the world. Language thus capitalises thought.

Touching, as is here only possible, but lightly on man's equipment in hand, and erect posture, and brain-power, perhaps enough has been said to show their interplay: brain, as controlling centre, guiding the organs to functions without which it was unable to develop and store up in its increasing folds and furrows the power which finally overcame the brute forces of nature.

For the quality of the brain in all animals is determined by the number of its furrows and creases. The brains of the lower vertebrates, as of fishes, are smooth-surfaced as well as small. The brain of a monkey is a sort of "skeleton map" of the brain of man; but when we reach that of the man-like ape we find "the details more and more filled in," while, in weight of brain, the difference between the savage and the civilised man is far greater than that between the savage and the highest ape. Man's brain-development is therefore due to his all-round activity; the nerves which ramify throughout the body conveying sensation to the "head centre," which, packed within its bony case, becomes more puckered and infolded as the nerves transmit their messages to its keeping. But brain, and nerve connections, and hand, were all of small avail, lacking tool and weapon; and it was from the moment when man made choice of a sharp-pointed flint to cut or kill something that his start in civilisation began. A little experience taught him the value of the hardest materials within reach, and there needed no higher degree of adaptive intelligence than some of the lower animals exhibit to shape the materials to his purpose. The beaver builds his log-house where neither flood nor foe can reach it, cuts long canals, and even makes locks where the stream-levels render the canals useless. The tailor-bird, with beak as needle, sews his nest of leaves with thread twisted of spiders' webs and cotton shreds; the wasp chews the wood to pulp whereof it makes its nest; the bower-bird builds a love-abode of sticks and shells, and flowers and feathers, where he and his mate may flirt and dance; and as for the ants—well, everybody knows what astounding perfection in their social life has been reached by creatures whose brains are, perhaps, more wonderful than the brain of a man; reached, too, ages before he appeared. But all these only repeat, never surpass, the skill of their ancestors.

Some explanation of man's probable place in time-history will now follow the account of his place in life-history.

CHAPTER II.

THE PLACE OF MAN IN THE EARTH'S TIME-HISTORY.

THE rocks which compose the crust of the globe are divided, speaking broadly, into two classes: Unstratified, or fire-fused; and Stratified, or water laid. The depth to which the former and older extend is unknown, but as they contain no remains of plants or animals, they tell us nothing about the order and relation of life-forms. The stratified rocks, which alone reveal that, are divided into the following epochs :—

	Estimated Depth.		Typical Plant.	Typical Animal.
Primary ...	136,000 ft., or	72.0	Seaweeds & Ferns.	Fishes.
Secondary...	25,000 ,,	13.4	Pines.	Reptiles.
Tertiary ...	27,000 ,,	14.6	Leaf-bearing trees.	Mammals.
Quaternary and Recent	500		Existing Species.	

The shell-bearing, soft-bodied animals, called mollusca, have been chosen as the most useful for the purpose of classifying the life-history of the globe, because they are more universally distributed through strata of every age than any other organisms. Hence they have been named the "alphabet of palæontology."

In accordance therewith the Tertiary Epoch is divided into three systems: the Eocene (Gr. *ēōs*, dawn, and *kainòs*, recent, or new); Miocene (Gr. *meiōn*, less), or less recent; and the Pliocene (Gr. *pleiōn*, more), or more recent. Sometimes the Pleistocene (Gr. *pleistòs*, most) otherwise, known as the Post-Tertiary, or Quaternary, system, is grouped with the Tertiary Epoch. The several terms indicate the relative percentages of shells found in each system.

All animals are probably vastly older than their earliest known fossil remains, and as far back as the Triassic, or oldest system of the

Secondary Epoch, we find relics of small mammals of the marsupial or pouched type, like opossums and kangaroos. But it is in Eocene deposits that remains of quadrupeds, representing most of the great mammals now existing, first appear. Both in the Old and New World — Europe and America being connected in Eocene times by way of Iceland and Greenland, and enjoying a tropical climate — there lived lemur-like creatures which had points of resemblance to the hoofed quadrupeds from which are descended the horse, rhinoceros, and other odd-toed animals; and the deer, swine, and other even-toed animals.. It was, therefore, not later than this period, when the several orders of mammals were in course of development, that the ancestors of lemurs, monkeys, anthropoid or man-like apes, and of man, appeared.

The earliest-known fossil anthropoid apes have been found in Miocene strata. In Northern India these have yielded remains of the chimpanzee, and in Western Europe remains of apes as large as man; the Dryopithecus (Gr. *drys*, an oak, and *pithēkos*, an ape), found near oak trunks at Saint Gaudrus, in Haute Garonne, and the Pliopithecus, an extinct gibbon, at Sansan, in Gers. Monkeys also, under a variety of forms, were present, and the representatives of living genera of mammals abundant. The land-connections of the northern hemispheres remained unbroken, and the climate, though lowering in tendency, so warm that water-lilies grew within eight degrees of the Pole, and a rich evergreen flora flourished throughout high latitudes.

Man is not, as has been shown already, the lineal descendent of his nearest relation, the ape, and it therefore follows that the division between the two cannot have been later than Miocene times. In fact, the evidence points to the divergence of the branch which includes monkeys and anthropoid apes, and of the branch which ends in man, about the close of the Eocene, or the beginning of the Miocene period, The deposits of the latter are almost a blank in Britain. But Miocene beds at Thenay, in France, have yielded a few relics of supreme interest in some flint flakes, which appear to be of artificial workmanship, even to bearing traces of splintering by fire. Were this last indication certain, their human origin would be undoubted. "Worked" flints have also been dug from like deposits at Puy Courny in Cantal, and near the Tagus, in Portugal, and there is full warrant for the expectation that, as Miocene strata are explored in other parts, corresponding relics will come to light. But up to the present time the various reports as to the discovery of such relics in those deposits have proved unfounded. The absence or scarcity of the earliest-known tokens of man's presence in one limited area are not evidence against his presence in other areas in Miocene times, and the rudely shaped flints of the Thenay and Tagus beds betray the "prentice hand" which must have preceded that of the expert workman. Man passed through many intermediate stages of long duration ere he became "the foremost in the files of time."

The Pliocene rocks tell us less about him than the Miocene. That is, so far as their contents have been examined—a very important qualification. They show great changes in this part of the globe in the sinking of the land between Norway and Iceland and between

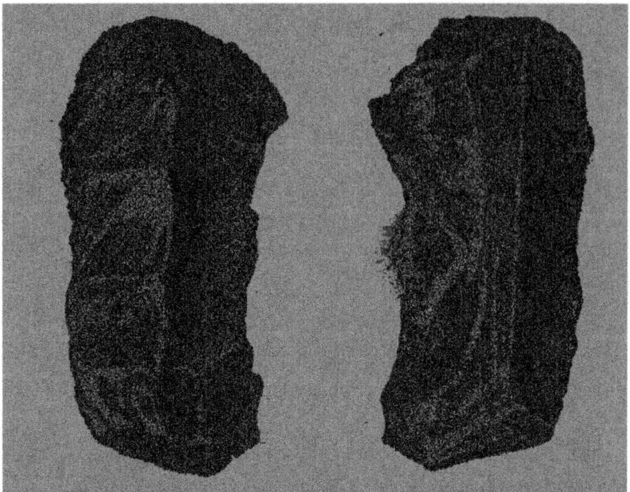

FIG. 3.—FLAKES FROM PLIOCENE BEDS, YENANGYOUNG, BURMA. (*Natural Science*, Nov., 1894.)

Britain and Greenland, thus sundering Europe from America; in the union of the waters of the Atlantic and Arctic oceans, and of these with the shallow North Sea, the area of which was extended. Although the climate was slowly growing colder, tropical plants, as the bamboo and pomegranate, flourished in Central Europe. Species of mammals now living were numerous: elephants and apes tenanted the forests, hippopotamuses wallowed in the swamps; the slow rhinoceros roamed the prairies, and the swift deer was the prey of hyænas. At the close of the Pliocene period the apes had disappeared from Europe, probably through the continued cooling of the climate, which resulted at last in the long reign of cold known as the Glacial Epoch, or the Great Ice Age, the causes of which—a change in the shape of the earth's orbit, and in the position of its axis—are beyond the province of this book to deal with. It must here suffice to point out the place of the Ice Age in geological history, and to add that it probably began 240,000 years ago, and, with intermittent periods of milder climates, came to an end 80,000 years ago; that during its greatest intensity it swathed the northern hemisphere in a winding-sheet of ice to the fiftieth parallel of latitude, dinting and rounding its surface, scoring the rocks with scratchings which they bear to this day, and sweeping away the northern flora, never to return. The same mighty agent left the world the poorer, from the life standpoint, in the destruction of the largest and strongest mammals.

The Glacial Periods—for there are at least three well-marked divisions—are included in the Pleistocene system. The plants and animals of the interglacial beds agree in all respects with those of non-glacial deposits, the interesting feature of both being the commingling of arctic and tropical forms, together with the undoubted evidences of man's presence associated with each. Animals, both extinct and living, inhabitants of widely separated areas, lived in the same region as the warmer and colder spells of climate alternated. The hippopotamus, rhinoceros, African lynx and elephant, lion, and hyæna—showing land connection between Europe and Africa—ranged as far north as Yorkshire; while the reindeer, Arctic fox, woolly-haired rhinoceros, and mammoth or woolly-haired elephant, roamed over Spain and Italy, and Scotch firs grew as far south as the banks of Lake Varese in Lombardy.

Among the extinct animals the huge mammoth deserves passing notice, if only from the association of its remains with relics of man in the valley of the Thames. Its name is derived from the Tâtar *mamma*, the earth, the natives believing that it lived underground, and that its burrowing was the cause of earthquakes. The Chinese have a legend that it died if it breathed the outer air. It had a wide distribution both in time and space, ranging over more than half the land-surface of the globe, and living from before the first stage of the Ice Age to the late Pleistocene period. In Siberia, its native home, where it existed in enormous numbers, not only have its large tusks, known as fossil ivory, been extensively used in commerce for centuries, but whole and perfectly preserved carcasses have been preserved in Nature's refrigerating chamber, the frozen layer of earth which underlies the surface soil of Siberia, extending not only beneath the treeless *tundra*, but also under forests and cornfields. In 1846, when the summer was unusually warm, a mammoth, standing upright in the place where it had been bogged countless ages ago, was thawed out of the icebound soil. It will have been noted that the materials for tracing the presence and movements of man, or of a creature who was "little less than" man, and "more than" ape, are too scanty and dubious to justify our saying positively that he lived in Western Europe before tht first Glacial Epoch. The question remains open. But, on the theory which all evidence confirms as to the divergence of manlike ape and apelike man in the late Eocene or early Miocene period, he must have lived somewhere. This brings us to the interesting question, In what part of the globe did man—*Homo sapiens*—originate? And to that question there is no answer; only an approach to one.

In the "Descent of Man" Darwin deals with the subject very briefly. He says, "It is probable that Africa was formerly inhabited by extinct apes closely allied to the gorilla and chimpanzee; and, as these two species are now man's nearest allies, it is somewhat more probable that our early progenitors lived on the African continent than elsewhere. But it is useless to speculate on this subject."

In an essay on the "Aryan Question" Huxley says that during the Pleistocene period "there is no reason to suppose that the genus *Homo* was confined to Europe; it is much more probable that this, like other mammalian genera of that period, was spread over a large extent of the surface of the globe. At that time, in fact, the climate of regions nearer the equator must have been far more favourable to the human species; and it is possible that, under such conditions, it may have attained a higher development than in the north. As to where the genus *Homo* originated, it is impossible to form even a probable guess. During the Miocene epoch one region of the present temperate zones would serve as well as another." ("Collected Essays," vii. p. 324.)

In "Darwinism," an "exposition of the theory of natural selection," of which he was the honoured and self-effacing co-originator, Wallace discusses the "probable birthplace of man" at greater length, but his views are too concisely stated to permit abbreviation, especially as they include theories on the obscure question of the rise of varieties or "races" of man. "It has usually been considered that the ancestral form of man originated in the tropics, where vegetation is most abundant and the climate most equable. But there are some important objections to this view. The anthropoid apes, as well as most of the monkey tribe, are essentially arboreal in their structure, whereas the great distinctive character of man is his special adaptation to terrestrial locomotion. We can hardly suppose, therefore, that he originated in a forest region, where fruits to be obtained by climbing are the chief vegetable food. It is more probable that he began his existence on the open plains or high plateaux of the temperate or sub-tropical zone, where the seeds of indigenous cereals and numerous herbivora, rodents, and game-birds, with fishes and molluscs in the lakes, rivers, and seas supplied him with an abundance of varied food. In such a region he would develop skill as a hunter, trapper, or fisherman, and later as a herdsman and cultivator.

In seeking to determine the particular areas in which his earliest traces are likely to be found, we are restricted to some portion of the eastern hemisphere, where alone the anthropoid apes exist, or have apparently ever existed. There is good reason to believe, also, that Africa must be excluded, because it is known to have been separated from the northern continent in early tertiary times, and to have acquired its existing fauna of the higher mammalia by a later union with that continent after the separation from it of Madagascar, an island which has preserved for us a sample, as it were, of the early African mammalian fauna, from which not only the anthropoid apes, but all the higher quadrumana (or 'four-handed," so called because the hind-feet can be used for grasping) are absent. There remains only the great Euro-Asiatic continent; and its enormous plateaux, extending from Persia right across Tibet and Siberia to Manchuria, afford an area some part or other of which probably offered suitable conditions in late Miocene or early Pliocene times (it will be seen that Mr. Wallace favours a more recent period than the Thenay and Indian flints appear to warrant) for the development of ancestral man. It is in this area that we still find that type—the Mongolian—which retains a colour of the skin midway between the black or brown-black of the negro, and the ruddy or olive-white of the Caucasian types, a colour which still prevails over all Northern Asia, over the American continents, and over much of Polynesia. From this primary tint arose under the influence of varied conditions, and probably in correlation with constitutional changes adapted to peculiar climates, the varied tints which still exist among mankind. If the reasoning by which this conclusion is reached be sound, and all the earlier stages of man's development from an animal form occurred in the area now indicated, we can better understand how it is that we have as yet met with no traces of the missing links, or even of man's existence, during late tertiary times, because no part of the world is so entirely unexplored by the geologist as this very region. The area in question is sufficiently extensive and varied to admit of primeval man having attained to a considerable population, and having developed his full human characteristics, both physical and mental, before there was any need for him to migrate beyond its limits.

In 1902, thirteen years after "Darwinism" was published, Dr. Dubois found in the upper Pliocene beds at Trinil, on the banks of the river Bengavan, in Java, a portion of skull, two molar teeth, and a thigh bone, which he assumed belonged to an animal named by him *Pithecanthropus Erectus*, or "upright ape-man." The forehead was low and narrow, the inner surface of the skull bore impressions of convolutions, and Dr. Dubois estimated that the brain was about twice as large as the brain of the largest anthropoid. Expert authority, represented by Professor Duckworth, is of opinion that we possess in these fragments (which occur in a region where the protohuman may have been developed) the nearest likeness yet found of the ancestor of man at a stage immediately antecedent to the definitely human phase, and yet at the same time in advance of the simian (or ape-like) stage.

We may now return to the sure ground as to man's presence in Europe in mid-Pleistocene times, and through all the climatal and other changes which preceded what may be more strictly called the Human Period. On the threshold of this we must stay to learn that the periods of time in Europe, from the unknown age of man's first appearance there, till about the Christian era, have been divided by Danish antiquaries into the "Ages" of *Stone*—sub-divided into the Palæolithic or Ancient Stone Age, and the Neolithic, or Newer Stone Age; of *Copper* or *Bronze*; and of *Iron*. This classification was anticipated by some ancient writers, notably by Lucretius in his great poem "De Rerum Natura," in the passage thus rendered by Mr. Munro.

"Arms of old were hands, nails, and teeth, and stones, and boughs broken off from the forests, and flame and fire, as soon as they had become known. Afterwards the force of iron and copper was discovered; and the use of copper was known before that of iron, as its nature is easier to work and it is found in greater quantity." (V. 1283–1288.)

In the peat deposits of Denmark, which range

from ten to thirty feet in depth, three distinct layers of trees occur. In the lowest beds there are found trunks of Scotch fir, which has not been a native within historical times, and no longer thrives when planted. Near them flint weapons, and bones of the stag and primitive ox, were discovered. At a higher level layers of oak were found, and with them some bronze shields, which are deposited in the noble Museum of Northern Antiquities at Copenhagen. In the uppermost beds trunks of the common beech, which still flourishes in Denmark, were found. Hence the classification.

By the *Stone Age*, as the term implies, is understood a period when metals were unknown, stone mainly, but also bone, shell, horn, wood, and such like accessible materials, being used by man as tools and weapons.

The implements of the *Ancient Stone Age* are alike in being of the rudest type, and neither ground nor polished; only roughly chipped. Some are found in the drift, or gravel-beds deposited by ancient rivers; others, of rather higher type, under the floors of limestone caverns. The implements of the *Newer Stone Age* are usually exposed upon or buried near the surface of the soil; they are found among the *débris* of rubbish-heaps or coastfinds, and lake-dwellings; and also in earthworks, tombs, &c. They are markedly distinct from palæolithic implements, not only in being fashioned of other stones besides flint, of which material the older implements for the most part are shaped, but in exhibiting manufacturing skill of a gradually higher order, and in being ground to an edge and more or less polished. Of course, chipping still went on. Very great differences in the features of Europe, evidencing changes stretching over vast periods, mark the two Ages, and appear to imply immense intervals between them, but the evidence for and against this is alike inconclusive. Upon the several stages in the shaping of stone implements the following quotation from Lubbock's "Pre-historic Times" is apposite.

"A very small step would lead man to the application of a sharp stone for cutting. When the edge became blunt the stone would be thrown away and another chosen, but after a while, accident, if not reflection, would show that a round stone would crack other stones, and thus the savage would learn to make sharp-edged stones for himself. At first, as we see in the drift specimens, these would be coarse and rough, but gradually the pieces chipped off would become smaller, the blows would be more cautiously and thoughtfully given, and at length it would be found that better work might be done by pressure than by blows. From pressure to polishing would again be but a small step. In making flint implements sparks would be produced; in polishing them it would not fail to be observed that they became hot, and in this way it is easy to see how the two methods of obtaining fire may have originated."

The implements of the *Copper* or *Bronze Age*, when a great advance in human progress was made possible by the discovery of metals, were, in the earlier period, fashioned of copper, afterwards, by the happy discovery of hardening it by mixing it with tin, of the compound bronze or "gunmetal."

The implements of the *Iron Age* were made of that yet harder and most valuable of all metals, which superseded bronze for cutting instruments, bronze being still used for the handles, and for ornamental purposes.

No definite dates can be given to these several divisions; the only thing certain is the succession of the stages of culture which they imperfectly denote. The Great Ice Age gives us a rough-and-ready measure of the vast time during which man inhabited Western Europe, in the Palæolithic Age. Compared with this, the Neolithic Age, so far as indicated by the time taken to form certain deposits in which its relics occur, is recent, say 5,000 B.C. The early Bronze Age may date about four thousand years later, merging into the Iron Age, references to that metal being found in the Homeric poems (about 850 B.C.) and in the "Works and Days" of Hesiod, the shepherd-poet, who describes the five ages of the world—gold, silver, bronze, heroes or demi-gods, and, lastly, iron, in which he lives, casting envious glances on the heroic times. But the exact figures have little importance. The several stages overlap, intermingle, and shade off one into the other "like the colours of the rainbow." They are not applicable to all parts of the world at one and the same time, as if there had been a universal abolition of stone tools and weapons at a certain period in human history, and a universal adoption of bronze tools and weapons in their place. It is highly probable that in the Ancient Stone Age the whole of Europe may have been inhabited by races using chipped and unpolished stone implements, but it is certain that in the later part of the Newer Stone Age it was occupied by races in very varying degrees of civilisation.

The more fortunate people who had settled in Southern Europe were, by contact with older peoples who sailed the Mediterranean, far ahead of those who were scattered in regions north of the Alps. Polished stone implements and, to some extent, bronze, which metal was exchanged in barter for the coveted amber washed on the Baltic shores, were in use among these long after iron was known to Greek and Roman. Widely as metals are distributed by commerce in the present day, barbarous people who make shift with stone tools and weapons still exist. But a few years ago there perished the last remnants of the aborigines of Tasmania, who in many respects nearly represented man of the Palæolithic Age. Further, the succession of the later ages is not universal, for some races, as in certain parts of Africa and Polynesia, passed direct from the use of stone to that of iron through the agency of traders. Attempts at exact chronological order are sometimes hindered by the retention of stone implements for ceremonial purposes. For example, the Egyptians, in embalming the bodies of their dead, made the first incision in the side of the corpse with a stone knife, and references occur in the Old Testament (Exodus iv. 25; Joshua v. 2) to the use of the same kind of instrument in the rite of circumcision, which, under certain conditions, is performed to this day with a fragment of flint or glass. The Brahman priest still kindles the sacrificial flame by the primitive method of producing fire—namely, rubbing two pieces of wood together till the sparks fly. In the most advanced times, in Mexico and Central America, the human sacrifice was slain with a stone knife on a stone slab, the neck and limbs being held down by a sacrificial collar and fetters of chased stone. These are a few specimens of abundant examples that, when the original purpose of a thing is forgotten or mystified, or when the use of it is restricted to a class, time and authority combine to invest it with sanctity. Dean Stanley has shown the operation of this in the case of certain vestments. "The *alb* is but the white shirt or tunic, still kept up in the white dress of the Pope, which used to be worn by every peasant next his skin, and in southern countries was often his only garment. The overcoat in the days of the Roman Empire, as in ours, was constantly changing its fashion and its name. One such overcoat was the *cape* or *cope*, also called '*pluviale*,' the 'waterproof.' Another was the *casula*, the 'little house,' as the Roman labourer called the smock frock in which he shut himself up when out at work in bad weather, and which survives in the *chasuble* wherewith the Roman Catholic priest decks himself before celebrating mass."

Our "swallow-tails" are only the old cut-away hunting coats, and the now purposeless buttons on the back were formerly used to fasten the skirts behind when riding. Man is much more a bundle of survivals than a "bundle of habits." All our pleasures and our pastimes are the outcome of primitive instincts and primitive practices. Our waltzes and quadrilles are the lineal descendants of barbaric religious dances; our conjuring is the comic offspring of tragic arts of sorcery; our plays and horse races are the representatives of dramas and games respectively, which were instituted as festivals in honour of the gods. Our picnics and campings-out satisfy a primitive nomad instinct; our fencing and boxing the old fighting instinct; and our "meets" the hunting instinct of remote ancestors who killed not for their pleasure so much as for their dinners.

CHAPTER III.

THE ANCIENT STONE AGE.

I. *Character of Remains found in the Drift.* The credit of the discovery of the meaning of certain objects found in old river-beds rests with a French savant, M. Boucher de Perthes. In 1839 he called the attention of other men of science to the finding of some rudely-shaped flint implements in hitherto undisturbed pits which were being worked for sand and gravel in the valley of the river Somme, near Abbeville, in

Picardy. They had been found at intervals during preceding years in such positions, and so far below the surface, as to convince him that they were of the same date as the deposits in which they were buried, and in which were also found bones of the mammoth, woolly-haired rhinoceros, and other extinct mammals. He argued that these "shaped" flints had been fashioned by man, and that they proved his great antiquity and low stage of civilisation. But, when he took his finds to Paris, he met with the same scepticism which the Abbé Bourgeois encountered thirty years later when he exhibited his Thenay flints in the same city. English antiquarians and geologists also looked askance at the Somme relics, but in 1858-59 their doubts were removed by a visit to the beds in which the implements were said to have been found. "In addition to being perfectly satisfied with the evidence adduced as to the nature of the discoveries, they had the crowning satisfaction of seeing one of the worked flints still *in situ* in its undisturbed matrix of gravel at a depth of 17 feet from the original surface of the ground."

All doubt being thus removed, an impetus was given to further research, and not only were discoveries of similar implements made in England in beds of gravel, sand, and clay, but it was found that flint implements—with never a thought as to their deep significance—had been unearthed many years ago; and, like the famous specimen from the Thames drift "opposite to black Mary's, near Grayes inn lane," put into a glass case, and labelled only "curious."

London and its neighbourhood, the old gravel beds of both Thames and Lea, have yielded abundant harvest—in short, "by 'many an ancient river' whose banks and course so largely determined the direction of man's wanderings, we find the tokens of his presence. We may take a run to the village of Caddington, near Dunstable, and noting the 'lakeside living place,' where Palæolithic man settled to work for a while, follow the course of the Lea to its junction with the Thames at Blackwall; gather chipped flints *en route* at Waltham, Edmonton, and Clapton, in short, all along the ancient clays and gravels, till we halt for a moment at Stoke Newington on 'an old surface floor which agrees well with the Palæolithic floor at Caddington.'"

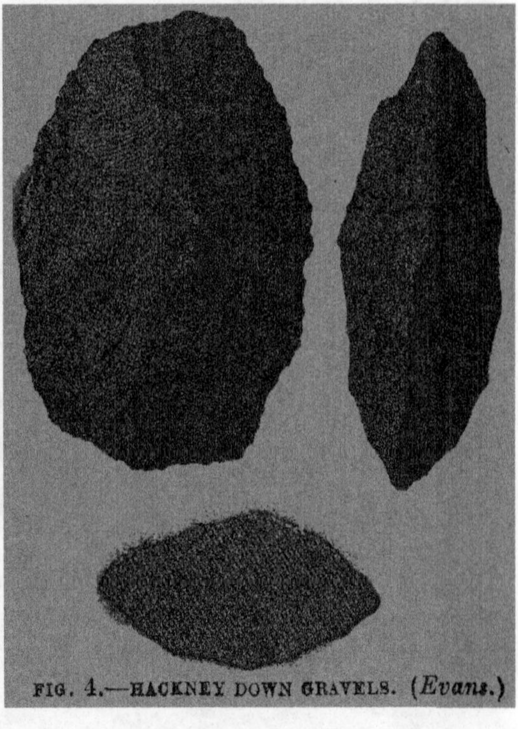

FIG. 4.—HACKNEY DOWN GRAVELS. (*Evans.*)

Sir John Evans, in his monograph on "Ancient

Stone Implements"—the standard work on the subject—divides the drift implements into three classes:

FIG. 5.—HIGHBURY NEW PARK. (*Evans.*)

1. Flint flakes, apparently intended for arrow-heads or knives. The durability of flint, and the ease with which, after practice, it can be chipped into the required form, caused it to be

more frequently used than any other stone. The flakes are removed either by blows or pressure, and were probably used by Palæolithic

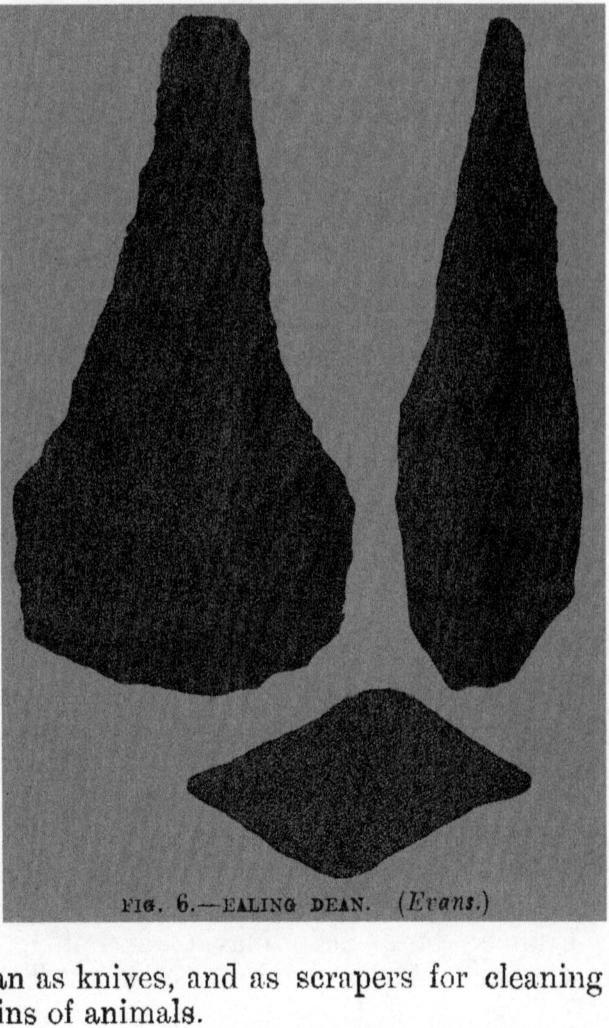

FIG. 6.—EALING DEAN. (*Evans.*)

man as knives, and as scrapers for cleaning the skins of animals.

2. Pointed weapons like lance or spear-heads.

3. Oval or tongue-shaped implements, presenting a cutting edge all round.

The manufacture of gun-flints for export to Africa is still carried on at Icklingham, in Suffolk, and at Brandon, in Norfolk. The workmen are few, and, with lessening demand for the flints, decreasing. Son has succeeded father through many generations, living the sombre, underground life which blanches their faces to the colour of their chalk surroundings. The production, attainable after a little practice, of stone implements with the aid of rough tools similar to those which the men of the Old Stone Age probably used, throws much light on the shaping of these earliest works of human skill.

FIG. 7.—FLAKE, PRESSIGNY. (*Evans.*)

But these cores and flakes and rudely-pointed missiles and tools are as like as two peas, and to further dilate upon them would invest these pages with the interest and charm of a catalogue. It suffices to say that there is a general resemblance of form between the implements found, not only in the river-drifts of southern England and France, but, Scandinavia excepted, of Europe, Asia, Africa, and America. "Their identity," as Boyd Dawkins remarks, "shows that the Palæolithic man who hunted the arnee (a variety of Indian buffalo) and the extinct hippopotamus in the forests of India; who wandered over Palestine and the valley of the Nile; who hunted the wild boar and stag, the mammoth, and, probably, the pigmy rhinoceros in the Mediterranean; was in the same rude state of civilisation as the hunter of the reindeer,

FIG. 8.—FLINT CORE WITH FLAKES REPLACED UPON IT. (*Evans.*)

bison, woolly rhinoceros, and horse in the forests of France and Britain."

It should be explained that the drift is formed of alluvial deposits—gravel, sand, clay, and stones — brought down by that slow yet ceaseless action of rain and flood which is for ever deepening the bed over which the waters flow. Since the time that the men of the Old Stone Age lived in France the Somme has scooped out its valley from 60 to 100 feet, a result which demands an enormous antiquity for the implements found in the gravels thus left high and dry, when we take into account the almost imperceptible rate at which our rivers are lowering their beds. For example, it is estimated that the Thames (apart from about 450,000 tons of chalk and other matter carried away annually in solution) lowers its basin at the rate of one foot in 11,700 years; the Boyne one foot in 6,700 years; the Forth one foot in 3,100 years; and the Tay one foot in 1,800 years.

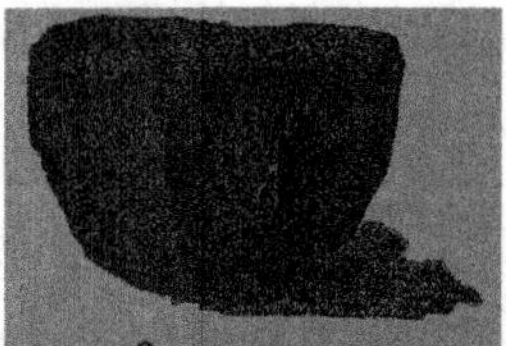

FIG. 9.—FLINT CORE, KENT'S CAVERN. (*Evans.*)

Britain was still part of the Continent when the relics of Palæolithic man fell to the bottom of the Somme. There were no Straits of Dover, and no English Channel. Lions prowled about the Mendips; droves of horses, herds of elk and reindeer, wild oxen, and smaller animals roamed over the plains now covered by the North Sea. Through these there ran a river fed by the several streams now known as the Rhine and the Elbe, the Tyne, the Humber, and the Thames. Each tributary, deepening its bed as it flowed, entombed therein the stone tools and weapons of man, and the bones of animals which the smacksmen on the Dogger Bank bring up in their nets in countless numbers. These fossil remains tell us what brute life surrounded man, and what varying climates prevailed.

II. *Character of Remains found in Caverns.* All changes which have taken place in the relative position of sea and land have occurred within the present continental areas; in other words, the deep basins which are filled by the great oceans have probably been as they now are from the earliest stages in the formation of the crust of the globe. But within the limits of change the havoc has been enormous; the rain, the atmosphere, and the levelling sea working destruction and wiping out traces of the past. One among the many evidences of this is that no caverns are earlier than mid-Pleistocene times. From immemorial ages, both to man and beast, overhanging cliffs and deep recesses in the hill and mountain-side have afforded "ready-made" shelters. And it is in caverns that, mingled with the bones of extinct and extant animals, further abundance of, and, compared with the river-drift, far more interesting, relics of man's presence have been found. Caverns generally occur in limestone rock, their formation being due to surface water from above, which, finding its way through some crack, and coming charged with carbonic acid derived from the atmosphere and from decayed vegetable matter, dissolves the limestone and slowly eats out a cavern. Then Nature, ever emptying the streams she fills, true to her cyclical action, oegins to choke up the hole she has made. The water, as it drips from the roof, deposits on the floor particles of lime held in solution, and these form the stalagmitic (Gr. *stalagmos*, a dropping) bed, which covers and hermetically seals-up any remains that may be lying about. The portions of dissolved limestone hanging, iciclelike, from the roof, gradually form the stalactitic columns which add to the beauty of many caverns.

A large proportion of the remains found in caverns have been swept-in by streams which once flowed at the same level as the cavern floor, the nature of the soil washed-in, and of the bones and other relics in it, giving the key to water action. Other causes explain the presence of objects which are so mixed as to confuse rather than instruct, but the greater number of accumulations can be accounted for only as due to the caverns having been used as the feeding-places and dwellings of man or animals. The fragments of food, and the tools and weapons scattered among them, witness to the use of the cavern by hunters, while in the crushed bones of small animals, commingled with the bones of great fleshfeeders, we find the traces of both eater and eaten.

The continued use of caves through long periods, and the frequent disturbances of the soil and its contents, render the period to which remains can be assigned less certain than in the case of those in the drift. But wherever bones of man and beast, or rude implements, are found together under lower beds of undisturbed stalagmite the fact of their high antiquity is assured.

As with the river-drift and its contents, so one cave is, speaking broadly, as good as another for the purpose of giving a common idea of the rest. Although many other caverns have been explored during recent years, the celebrated example known as Kent's Hole near Torquay remains one of the best specimens of its kind. Entering that ancient haunt of man, we find the deposits in this order, beginning with the uppermost:

1. Blocks of limestone, weighing from a few pounds to upwards of one hundred tons, which have fallen from the roof at various times, and become more or less cemented by carbonate of lime.

2. A black muddy mould from three inches to

FIG. 10.—HAMMER-STONE, KENT'S CAVERN. (*Evans.*)

FIG. 11.—BONE NEEDLE, KENT'S CAVERN. (*Evans.*)

a foot in thickness, composed almost entirely of decayed vegetable matter. This is known as the Black Mould.

3. A floor of stalagmite of granular character, varying in thickness from three inches to upwards of five feet.

4. A layer composed mainly of charred wood. This is about four inches in depth and occurs only in one part of the cavern. It is called the Black Band.

5. A light red loam, called the Cave Earth.

6. Another stalagmitic floor, but of a crystalline character, twelve feet thick in some places.

7. Below all these is a dark-red sandy deposit free from limestone, called breccia.

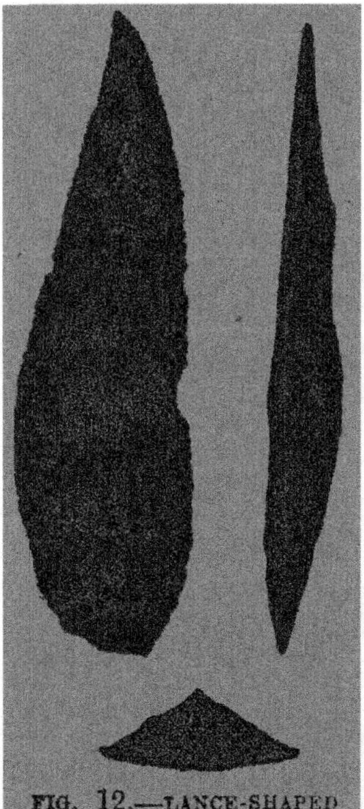

FIG. 12.—LANCE-SHAPED INSTRUMENT, KENT'S CAVERN. (*Evans.*)

The objects found in the black mould are, comparatively speaking, modern, although belonging to different periods, such as bronze knives, fragments of roughly smelted copper, and Roman or pre-Roman pottery, but associated with stone and bone implements. The upper bed of stalagmite, the black band, and the cave earth yielded remains of a very mixed group of animals: the mammoth, woolly rhinoceros, cave lion, cave bear, reindeer, Irish elk, horse, &c., and numerous flint flakes and cores, *i.e.*, nuclei or remnants of flints from which flakes have been struck off by a blow or by pressure. Besides flint tools and flakes, the black band contained a bone awl, bone harpoon and needle with well-formed eye. In the caves of the Dordogne, in France, the stone implements with which the eyes were drilled in the bone needles were found. The cave earth was richest of all the layers in remains; the lower stalagmite contained only bones of the fierce cave bear, and in the breccia there were found associated with remains of that animal (which included a "worked" fossil tooth) implements of flint and chert of much rougher type than those in the cave earth. These several accumulations represent an enormous antiquity for cave-man in Western Europe. Until seventy years ago, none of them had been disturbed since the time of their slow deposition; slow, because a stalagmitic floor cannot be formed quicker than the limestone overhead is dissolved, and the rate of that dissolution depends mainly on the amount of carbonic acid in the water. Air currents and other causes also affect the rate of deposition. The scribbling of one's name on public monuments and elsewhere, presumably that the "world" may know the momentous fact of our having visited such and such a place, is a senseless disfigurement of things, but, very rarely, it renders unwittingly a public service.

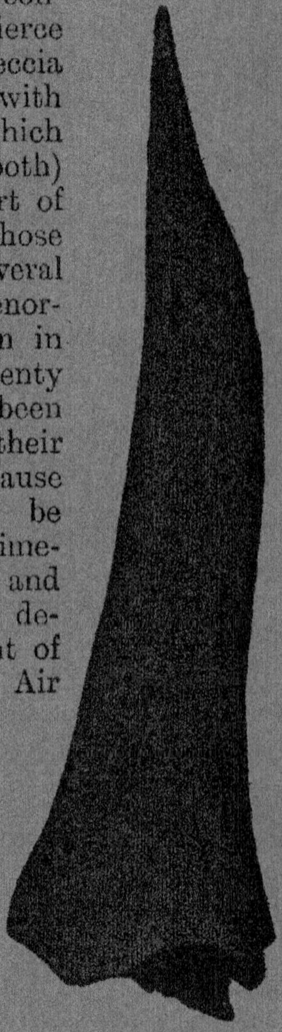

FIG. 13.—BONE AWL, KENT'S CAVERN. (*Evans.*)

This a certain "Robert Hedges of Ireland" did on February 20th in the year of Revolution, 1688. For, entering Kent's Hole, he cut his name on a boss of stalagmite, and a Roman Catholic clergyman, the Rev. J. MacEnery, one of the first explorers of the cavern, has left on record a description of the appearance of the letters in 1825. He says that they are "glazed over and partly effaced," a description which still applies, although the water has been depositing carbonate of limestone on the boss—it being vertically beneath a stalactite—for just upon seventy years since his visit. The film which has thus ac-

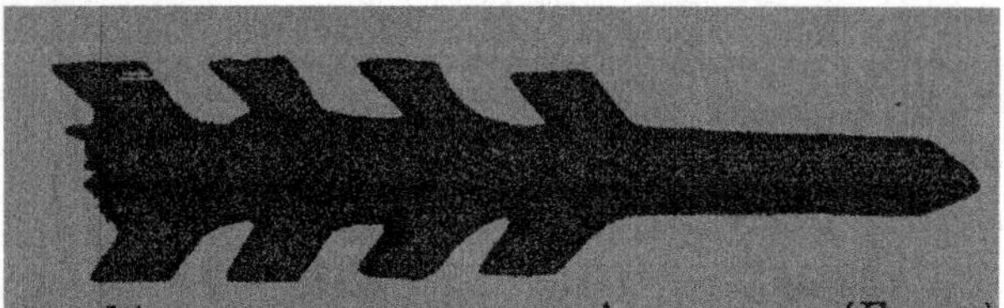

FIG. 14.—HARPOON HEAD, KENT'S CAVERN. (*Evans.*)

creted in two centuries is about the one-twentieth of an inch in thickness. Assuming an even rate of deposit, the time demanded for the accumulation of the two layers of stalagmite, to say nothing of the cave earth between them and the breccia or older cave earth below them, is astounding. But it must suffice to convey the impression of high antiquity, and leave out figures, especially as a similar boss in the Ingleborough cave in Yorkshire has grown at the rapid rate of ·2941 inch per annum, showing that thickness is not to be invariably taken as the measure of time.

The likeness in general type of the implements found in the lowest cave deposits to those of the river gravels evidences that the caves were used as shelters by the drift-men. Then, after intervals whose duration is explained by the upper deposits just referred to, the cave-men appear. Probably they were descendants of the drift-men, and they had certainly reached a somewhat higher state of culture. This is shown in the more finished workmanship of their tools and weapons; in the greater variety, both of materials and the application of them, and in traces of the arts of life which occur in the cave deposits. As distinguished from the drift-men, who are identified with the Mammoth period, they are often spoken of as men of the Reindeer period, because, although the times in which both animals appear overlap, the reindeer, as the remains of the hunting feasts show, was one of their chief sources of food.

The range of the Palæolithic cave-dwellers—of course not necessarily during the same periods—was as wide as the habitable globe. Traces of their presence occur in caverns from Yorkshire to Gibraltar, from France to Syria, and across the Pacific to America. But caverns are found only in limited areas; and in the shelterless open country and the beast-haunted forest man made his hut of earth or boughs, or dug his pit for refuge. Along the river-valleys of Western Europe and elsewhere great heaps of refuse mark the sites of fugitive settlements of which all other traces have been long swept away.

The Abbé Bourgeois argued that the worked flints found in the Miocene beds at Thenay bore traces of having been fractured by the aid of fire, or used as "pot-boilers" for cooking. But the evidence is not conclusive, although supported by corresponding finds in deposits of sand near Orleans, because the flints may have been fractured by lightning or other natural causes. Less doubtful is the witness of some calcined stones in the gravels of Ealing, and certainty attaches to the knowledge of fire among the cave-men in the ashes, cinders, charred remains, and wholly or partly burnt bones amongst the *débris*. It would have been surprising if these people, who had certainly reached a stage of culture little, if at all, lower than the Fuegians and Tasmanians, had been ignorant of an art—one of the oldest inventions of man—with which no race of savages is known to be unacquainted. The original source whence man obtained fire, or the occasion which suggested its production, have given rise to a mass of myth and legend among the unlearned, and to much speculation among the learned. It seemed to the untutored mind that so mighty an agent in the progress of man—without which, indeed, no advance beyond the barbaric state was possible—must either be the gift of the gods, or have been stolen from them by some daring hero—one of the great culture-heroes to whom all races trace their arts and civilisation. Mysterious alike in nature and in origin, no wonder that it became the object of widespread worship; symbol of the divine, as among the so-called "fire-worshippers," the Parsis; and the mystic element whose guardianship was the prerogative of a sacerdotal caste—vestal virgin of Rome; Brahmin; priestess of Peru; or priest of Baal. Reference has been made to the Brahmanic production of the sacred flame by primitive methods, and if, through any neglect of a vestal priestess, the fire in the temple went out, a new flame was kindled by friction on a consecrated piece of wood. Friction was, of course, the earliest mode of its production, either by the laborious rubbing of two pieces of dry wood together, or by the more commonly used fire-drill, consisting of a stick placed in a cavity and twirled rapidly between the two hands. All other modes of procuring fire are but forms of friction based on the observation that motion develops heat. Many things would show this, as, *e.g.*, the drilling of a hole in a stone, the emission of sparks from a stone when struck, the striking of a tree by lightning, the rubbing of its branches together, or the personal sense of warmth in rubbing the hands.

As the cave-men had no pottery—such remains as occur belonging to the Neolithic Age—they must have cooked their meat in savage fashion; that is, either by putting it on a rough spit, or direct on glowing embers or red-hot stones, or by dropping the stones into water poured into stone cavities, or into holes lined with clay or hide, and then popping-in the meat when the water boiled. This last was the method among the Red Indians or "stone-cookers," as they were called, before the traders supplied them with earthenware vessels. About three centuries ago heated stones were used in Ireland for warm ing milk, and within the same period meat was cooked in the skin of the animal in the Hebrides. The Polynesians wrapped the meat in leaves and put it on heated stones in a pit, a method which Mr. Romilly saw in practice at a cannibal feast in New Ireland ten years ago.

The use of calabashes, coconut-shells, and of any other hollow natural object, as the skull or horn of an animal, for drinking purposes, is obvious. The word "keramic," applied to all fictile ware, comes from the Greek *keras*, which shows the use of horns for drinking-cups, evidencing what was their original shape and nature. There is little doubt that the invention of pottery was due to the practice of coating the outside or inside of inflammable vessels, as bowls and baskets, gourd rinds, and so on, with clay to protect them from fire. When it was found that the clay not only stood the heat, but was baked hard by it, the material which it covered was discarded as needless. The most primitive ornamentation on pottery is made with the finger-nail, and is often a rude imitation of the traces left by the basket-work or rushes.

Considering their rude weapons, the cave-men were "mighty hunters;" for the mammoth, woolly-haired rhinoceros, and other huge beasts, were their quarry. The reindeer's antlers were convenient daggers, the large pear-shaped flints powerful missiles, and the spears, tipped with sharply pointed stone or bone, were deadly darts. Small three-cornered flints from the drift—rough drafts of the exquisitely made arrow-heads of the Neolithic age—imply knowledge of the bow

FIG. 15.—SAVAGE ORNAMENTATION.

in early Palæolithic times; the barbed arrows, or spear-heads, so commonly found in the caves of France being used for fowling, and bone harpoons and barbed hooks for fishing.

While the men were at the chase the women "kept house," cooked the food, and made the clothing. This consisted of skins of the hunted animals—for as yet none were domesticated—sewn together by bone needles with threads of sinews or intestines. Passing by other dry and less needful details in the filling-in of a rough sketch, we come to a matter of deeper human interest in the genuine relics of primitive art found in the caves of Périgord and elsewhere in France, Belgium, and Switzerland. Here the Reindeer men have left "more vivid pictures of their life and times than those founded upon implements and weapons and the associated animal remains. Fortunately for us they employed the intervals of leisure from the chase in engraving upon bone, antler, and, more rarely, on ivory and stone, the hunting scenes which most vividly impressed themselves upon their memory." It is in the caves of the Dordogne that the most remarkable of these examples have been found. Grouping them together, without precise reference to place, we have in one sketch, cut on a piece of antler, a wild ox (*urus*) feeding, while behind him is a creeping man in the act of throwing a spear. In another, a naked hunter is also hurling a spear at a horse; another shows a group of reindeer, of which two are walking and three—probably captured—are lying on their backs. In another, representing

FIG. 16.—HORSES INCISED ON ANTLER, LA MADELAINE CAVE IN THE DORDOGNE.

an ibex, "the fragment of antler on which it is engraved was probably broken after the artist had begun his work, without leaving room for the completion of the figure. But the proportions are not sacrificed, nor is the animal deprived of its hind legs, which are doubled forward until they touch the under surface of the body." One

FIG. 17.—GROUP OF REINDEER; SCRATCHED ON SLATE, LA MADELAINE.

of the most spirited examples is the well-known figure of a mammoth scratched on a fragment of ivory, in which the artist has faithfully drawn the animal's shaggy ears, long hair, and upwardly curved tusks, concealing its feet in the high grass which covered them. The preservation of entire carcasses of this creature, already referred to, enables the accuracy of the picture to be verified. The list of specimens, in which the reindeer is the animal most frequently represented, might be considerably lengthened, but it suffices to add that the discovery of the figure of a horse on a small fragment of rib in the Robin Hood cavern has "high value in bringing the cave-men of Britain into relation with those of France, Belgium, and Switzerland."

Nothing has yet been said about the remains

FIG. 18.—SKETCH OF MAMMOTH ON FRAGMENT OF IVORY, LA MADELAINE.

of man himself in skulls and other parts of his skeleton. Compared with his imperishable works in flint and such like substances, these are extremely rare. For this there are sufficing causes. There is the fact, to which Sir John Lubbock refers, that in the gravel-beds of St. Acheul, near Amiens, "*no trace has ever been found of any animal as small as man.*" The larger and more solid bones of the elephant and rhinoceros, the ox, horse, and stag, remain, but every vestige of the smaller bones has perished. Not only were Palæolithic men widely scattered; their numbers, relatively to other animals, were small. Basing these on estimates of the proportions among hunting tribes, the figures would be about 750 to 1; and allowing for the length of man's life as 4 to 1, it follows that about three thousand skeletons of different animals of the chase would be left for one human skeleton. And of man's bones the hyænas would make short work. Then there is the dissolving action of certain acids, especially in peat, to be taken into account; the floating of bodies to the sea; the small area of ground yet opened up in which human bones, or the less perishable teeth, may be imbedded; and, not to cite more reasons, the areas once occupied by man, but now submerged, and therefore inaccessible to research. As bearing on the subject, there is the curious result of the draining of the Lake of Haarlem some forty years ago. Although a large population had lived on its banks, and although vessels had been wrecked in it, and naval battles fought on it, the engineers found no human bones whatever in deposits which had constituted the bed of the great lake for three centuries.

So many of the assumed finds have occurred in deposits the period or disturbance of which is not beyond question, that reference will be made only to a few of the discoveries which are less disputed. The river-gravels of the Continent have, as yet, yielded no skeleton of man of the drift; and those of Great Britain are alike barren. Such remains as have been recovered are found in the lower deposits of caves, of the use of which, as places of sepulture from early to later times, there are abundant traces.

Among the most important discoveries was that yielded by the cave of Duruthy, in the Western Pyrenees, where a crushed human skull and some scattered finger-bones were found associated with the rudest types of flint implements—flakes and scrapers—all imbedded in a hitherto undisturbed mass, above which was a sepulchral chamber containing numerous skeletons of the Newer Stone Age. Near the Palæolithic skull was a number of perforated teeth of bears and lions, lying in such a manner as to prove that they had formed part of a necklace. Moreover, the teeth are scratched with ornamental designs—barbed harpoons, arrowheads, and the figures of a pike and eel, and of a pair of gloves. For man early betrayed the love of ornament; and the rouge pot, in the shape of oxide of iron, as well as the necklace of shells or teeth, have been found in cave deposits. The Papuan, who swallows dirt and weapons, and decorates himself with coloured berries; the dancing Feejeeans (albeit converted to Wesleyism), who painted one half of the face red, and the other half black; the Admiralty Island natives, who were delighted at being covered with stripes of yellow and green paint when they went on board the *Challenger;* and the Esquimaux, who wear a stud in the lower lip or cheeks, are one with the modern dandy and the powdered beauty of to-day in æsthetic descent from the ornamented dwellers in Palæolithic caves.

As to other skulls, one found at Canstadt, near Stuttgart, in 1700, but not examined till 135 years later, was pronounced to be that of a

FIG. 19.—TATTOO ON A MAORI'S FACE.

Palæolithic savage. Twenty years before the Duruthy bones came to light, skulls had been found in the Neanderthal cave, in Germany, and in a cave at Engis, in Belgium. Bones of the mammoth and woolly rhinoceros were associated with the Engis cranium, pointing to its great antiquity; but, concerning both it and the

Neanderthal specimen, Huxley says that neither of them fills up or lessens the structural interval between man and the man-like apes. Which is exactly what may be expected, since, as shown already, the divergence between man and ape occurred at a period remote enough to bring about the differences which mark the one from the other. The Engis skull is a fair average specimen; "it might have belonged to a philosopher, or might have contained the thoughtless brains of a savage." The Neanderthal bones demonstrate the existence of a man whose skull may be said to revert somewhat towards the pithecoid (ape-like) type. Most of us, probably, have met people of whose heads the same might be said.

If conclusive evidence were wanted, the two skeletons found in a cave in Spy, in Namur, in 1886, appear to supply it. Upon these their discoverers reported that although they possess a greater number of ape-like characters than any other race of mankind, "between them and an existing anthropoid ape there lies an abyss." They add that "the distance which separates the man of Spy from the modern anthropoid ape is undoubtedly enormous; between the man of Spy and the *Dryopithecus* it is a little less. But we must be permitted to point out that if the man of the later Quaternary age is the stock whence existing races have sprung, he has travelled a very long way. From the data now obtained it is permissible to believe that we shall be able to pursue the ancestral type of men and the anthropoid apes still farther, perhaps as far as the Eocene, and even beyond."

Before we open another chapter of this history it may be well to focus what has been said about the races of the Drift and Cave period, and attempt some picture of them from the vague and scattered materials we possess. Help thereto will come from knowledge of the condition of savage peoples who are still, or were quite recently, in the Stone Age.

It is needful to bear in mind that the term "primitive" as applied to man, and, indeed, as applied to all higher forms of life, has no scientific accuracy; and is used only for convenience as denoting the highest point which our knowledge about the type described has reached. The man of the river-drift, as has been shown, was the descendant of a yet more primitive form.

First, then, as to his body; next, as to his mental faculties; and, lastly, as to his social life.

1. Taking the skeletons of Spy as types, Palæolithic man was powerfully built, although of short or stunted stature, probably about five feet, like the Fuegians, Bushmen, Mincopies of the Andaman Islands, and other extant savages. Broad-legged, with curved thigh-bones, his walk was shambling, as that of the gorilla or of bandy-legged persons. His long skull had a low, receding forehead with overhanging brows, furnished with bushy hair; the nose was flat, the nostrils large; the ears somewhat pointed, the big heavy jaw "prognathous" or "snouty"; the canine teeth fang-like, and the chin very small and retreating. The skin was probably

copper-coloured, and largely covered with long, straight hair like that of the Ainu of Yezo, the northern island of Japan. If the females differed at all from the males it was probably only in being of rather shorter stature.

2. Mentally, the "ape and tiger" were but little subdued in him. His feelings were rarely under control; the impulse of the moment ruled his life. "The mind of the child with the strength and passion of the man" were blended in him. Cunning he was, because he had to live by his wits; to kill and, probably, eat his foes, if he would not be killed and eaten by them; to fight without pause for food for himself, for the mate whom he had won, and for the child that she had borne him; the common need and common peril strengthening the social life which began in a remoter past. This struggle involved the constant exercise of the senses; hence the sharpening of sight and hearing, so that he could see and hear things to which the civilised man, dulled by artificial aid and by less need for alertness, is both deaf and blind. The earth was a telephone, to which, instinctively using the "method of Zadig," he put his ear and listened to the distant tramp of his enemy. With unerring skill he could with a stone missile bring down the bird on the wing; transfix the fleeting prey with his flint or bone-tipped spear; and, diving into the water, bring up the fish with a finger in each eye, like the South Sea Islanders or the Australians, who will dive, spear in hand, and come up with a transfixed fish. Living only for the day, he had no

thought for a morrow which might bring starvation. Beyond his tools and weapons—and these often lost or cast away—he had no possessions to which to cling; for the wandering hunter has no hearth to protect. Outside the little family or group there was no pity or sympathy; because the enlargement of these comes only as the social life widens.

With reasoning faculty but little developed and centred on bodily needs, such ideas as things around suggested to his twilight mind were a tangle of confusion, contradiction, and bewilderment. As he had but the dimmest idea of the relation of one thing to another, he could not group them under general facts. He knew nothing of the causal connection between a person or thing and its shadow; between sleep and dreams; between a cloud and its reflection in the water; between a sound and its echo from the hillside. He saw that the sun and stars came and went, that the water or big rivers fell and rose, or in the case of smaller streams sometimes disappeared altogether, and, falling from the sky, filled them again. In these and a hundred other events he dimly noted the differences which, in the long run, lead the mind to comparisons, and thereby lay the foundation of knowledge—of the relations between things which we call cause and effect. But to bring out of this the conception of law and order needs more than the experience of one life; ages passed before man could correct the first impressions of his senses and learn the facts about his surroundings. "If," as Pfleiderer says, "we require whole years to develop abstract ideas in the minds of our children, though they have the benefit of all their inheritance from the past, 'which thought for them,' it must have required centuries, and even millenniums, for primitive man to arrive at the same results."

Of course at the lowest stage that we can put him, "thinking without knowing that he thought," he was picking up knowledge for the advantage of all who came after him. Knowledge of the haunts and habits of the prey which he sought; of the fittest seeds and berries for food; of the times they fruited, and of the soil in which they grew; of the elemental differences in things, as the sinking of stone and the floating of wood; of the properties of things, their hardness or softness; their sweetness or bitterness; of the strangeness of things, as when the struck flint emitted sparks that made him think fire dwelt inside it, or that it was alive. Eye and ear and brain, thus kept alert, fed the sense of wonder which took the oddest shapes, to know which gives us the key to those workings of the primitive mind in which lie the beginnings of science and religion; the slow passage from guesses to certainties. For, at the start, man was befooled by his senses, and it has taken him, at a cost that makes the thoughtful weep, thousands of years to escape from the false impressions of things which they conveyed. His eyes told him that the earth is flat and fixed, and covered in by a dome-like vault, across which sun, moon, and stars pass. His ears told him that what we know to be the echo of our voice was made by mocking spirits, who also howled in the wind and roared in the thunder—spirits with which his imagination, ruled by his fears, peopled everything. For in the degree that he was able to reason at all, or to compare one thing with another, he saw seeming likenesses in things most unlike, and so was led into all sorts of pitfalls of the mind. Because he moved, he looked upon every moving thing as alive like himself. Rustling leaves, waving grass, rolling stone, swirling water, drifting cloud, rising and setting bodies of the sky, all, to his thinking, were alive, and full of passions and feelings as he was. Or, if not alive themselves, they were controlled by some life. Hence arose belief in spirits everywhere; at the first, baleful and malignant, because in the degree that the nature of a thing is unknown or misapprehended, it is dreaded. Knowledge, like love, casts out fear. And since fear always magnifies the supposed power of that of which we are afraid, it is easy to see how stones and trees, water and stars, and a heap of other inanimate things, came to have offerings and sacrifices made to them to appease their anger or win their favour. So we may say that with belief in spirits arises savage religion, and that in guesses about things—real enough to their framers—arises savage science; the religion and the science being entangled and mixed together in the primitive mind.

In what has been now said we may seem to have travelled a long way from the mental standpoint of man in the Old Stone Age. For he was certainly not in advance of the larger number of modern savages whom all travellers

FIG. 20.—STONE CLUB, NEW BRITAIN. (*Powell.*)

agree in reporting not only as listless and incurious, but as unable to fix their attention for even a few moments on anything out of the common. The difficulty in all attempts to define the mental power of men so low in the scale is that we cannot put ourselves in the place of people whose language is made up of jabbers and gestures, who have the vaguest idea of a to-morrow, and who cannot count beyond three. But the power in man to develop into what the highest specimens of his kind have become was in him at his lowest, and it is needful to keep in mind that we are dealing with a series of mental stages in which there is no break, but at the lower of which man remained for an enormous period.

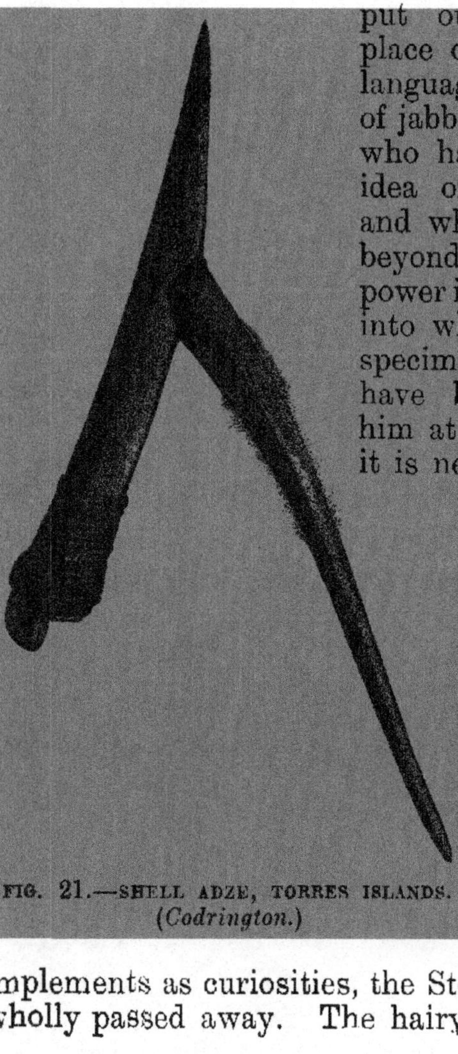

FIG. 21.—SHELL ADZE, TORRES ISLANDS. (*Codrington.*)

3. Although travellers nowadays buy stone implements as curiosities, the Stone Age has not wholly passed away. The hairy Ainu who, like

the filthy Hottentots, never wash themselves from birth till death, use bone and bamboo arrow-points in hunting and fishing, and live on raw flesh, seaweeds and roots. They have no marriage customs, a man taking as many wives as he can afford. The natives of New Britain, until the quite recent introduction of iron by white traders, used stone tomahawks wedged between two pieces of wood, and tipped their spears with the bone of a slain enemy so that his power might be added to their own in hurling the weapon. The cannibal races of Queensland use wood for most of their weapons, but have tomahawks of basalt or other hard stone, using sharp-edged implements to rip open the carcase of the animals killed. They also eat beetles, grubs, and vermin. Amongst the Melanesians the adzes on one group of islands are of stone, on the other group they are made of the giant clam shell: their spears are armed with bone points. The Nicaraguan Indians fix their stone hatchets in

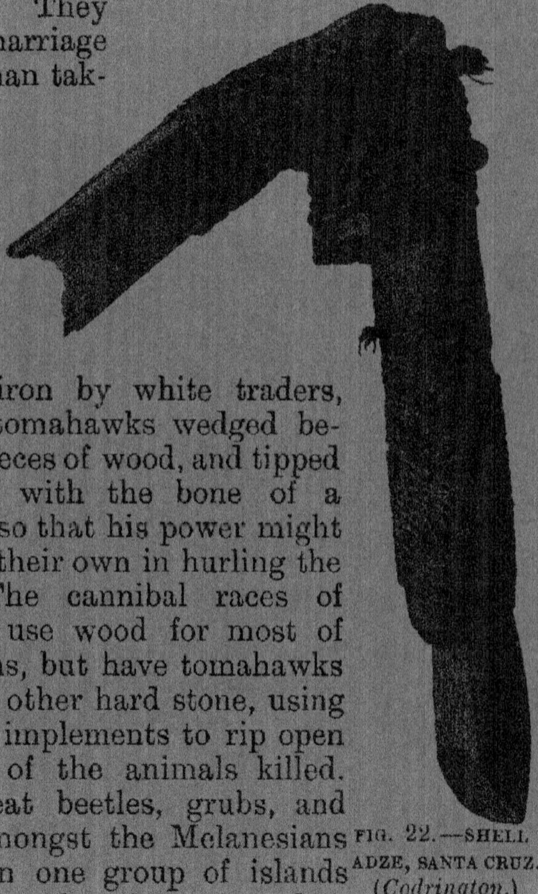

FIG. 22.—SHELL ADZE, SANTA CRUZ. (*Codrington.*)

stone-cut wooden handles. The Tasmanians, who in many respects most nearly represented the presumed general condition of Palæolithic man, and the extinction of whom is a reproach to "civilised" people, used weapons of chert of the rudest make, which they grasped with their

FIG. 23.—NATIVE HOUSE, TESTE ISLAND, NEW GUINEA. (*Powell.*)

hands, being ignorant of the mode of hafting them. Their canoes were a float or raft of bark bundles which was propelled by a pole; they lived under bough shelters, and made fire with the simplest and perhaps oldest of all inventions, the fire-drill. They drew rude pictures on bark, were quick and cunning in their own sphere, but stupid outside it. In their crude religious ideas they conceived of the shadow of anything as its ghost; the echo was the "talking shadow," and they believed in evil spirits. They buried their dead and avoided their graves, a custom

FIG. 24.—ESQUIMAUX WINTER HUT (NATIVE DRAWING).
(*From Rink's Tales of the Esquimaux.*)

which, as will be shown presently, indicates fear of the ghost.

These few examples, taken haphazard, might be extended, but they supply sufficing material to fill up some gaps in our sketch. Man of the Ancient Stone Age was in the hunter stage of culture, but without any domesticated animal as his help. His temporary homes depended upon seasons and places: the Veddahs of Ceylon make their huts of boughs and bark; the Hottentots use sticks and mats; the Esquimaux in summer-time stretch skins upon bones lashed together to make posts; in winter they build huts of wood or drift timber; the wretched Fuegians sleep on the unsheltered ground, and the Bushmen, when away from their caves on hunting excursions, bury themselves in the sand. Broadly speaking; for the nomad, the tent or the cave, from either of which he shifts easily; for the tiller, the settled homestead. As with shelter, so with clothing; climate and zone rule that. Nakedness is not necessarily immodesty, and the gift of clothing to a savage people has sometimes been their ruin. Different races cover different parts of the body; in the East women conceal their faces, and a strict Moslem would be shocked at the bare neck and shoulders of Western women in evening dress. The simplest form of clothing among natives of

FIG. 25.—AN AUSTRALIAN GRAVESTONE.

FIG. 26.—BUSHMAN WALL-PAINTING.

warm climates consists of leaves or twigs, or pendant strips or fringes round a girdle; in colder climates the skins of slain animals are the primitive dress. The delight of the savage in a painted or otherwise decorated skin has been already referred to. And this leads to the interesting parallels to the art of the cave-men furnished by other savages than the Tasmanians. The "black fellows" of Australia rank low in the scale, but they have depicted sharks, porpoises, lizards, weapons, and canoes on the faces of rocks; and on "grave" pillars have sketched the *boilyas*, or ghosts of dead men and animals. The Bushmen have also painted figures in red, brown, and other colours on cliffs, or etched them in light tints on a dark ground; and the drawings of North American tribes on stone and bark rank high in savage art.

The beginnings of social life go back to a time before man and monkey had branched off from their common stem, and whether or not our pre-human ancestor had a special pairing season, man, as we know him, paired at all seasons of the year, and remained faithful to his mate as food-winner and protector, at least during the infancy of the offspring. Even the anthropoid apes do that, and in man, as the germs of sympathy on which family life depends developed, and as the period of infancy of the offspring was lengthened, there was cultivated the deeper social feeling. Thus loose and fitful relations tended to become lasting. But that advance was slow; among a wandering hunting tribe "useless mouths" are an encumbrance, consequently infanticide, especially of females and

FIG. 27.—PAINTINGS ON A CROW (NORTH AMERICAN INDIAN) ROBE. (*Catlin.*)

puny or sick children, was largely practised, and

a blow with a stone axe settled the fate of many an aged burden, the dead being probably left to be devoured by hyænas and other wild beasts. There is no trustworthy clue to the mode of disposal of his dead by Palæolithic man; the relics of funeral feasts which point to cannibalism—broken skulls and human bones split to extract the marrow—are in early neolithic deposits; and perhaps he was on a level with the cave-dwellers by the Red Sea, of whom Diodorus Siculus tells as "mocking at all manner of sepultures, for as soon as any of them is dead, they tie his head betweene his legs with a withe of hawthorne or willow, and dragging the corpse to the highest place they can finde, with laughter and jeering, they overwhelme it with stones, and then putting a goat's horn on the top of the stones, they leave it there without any pitty or compassion at all."

The weeding-out process which man, in all stages of civilisation—whether savage of the Stone Age, barbaric Gaul, cultivated Greek or Roman—has carried out, and which still prevails among a large portion of the human race, has been aided by the continuous action of "natural selection." That action, it is almost needless to say, is involved in the tendency of all species to multiply beyond the means of subsistence; and in the variations, for the most part slight, of offspring from their parents. The first cause gives rise to ceaseless struggle for existence among *all* living things, for, as Darwin points out, "even slow-breeding man has doubled in twenty-five years, and at this rate in less than a thousand years there would literally not be standing room for his progeny." And in this struggle the destructive agencies of nature intervene. They who win in the merciless competition do so in virtue of some favourable variation which the vanquished lack; for the race is to the swift, and the victory to the strong. But exception to the unchecked action of natural selection arises in man at a certain stage—necessarily a high one—of his development as a social being. His association into civilised groups enlarges the sympathetic feelings and, as one among other results, brings to no mean state of perfection, faculties, as the mathematical and musical, the development of which is not due to "natural selection," or to the struggle between man and man. But this matter, apart from the obscurity which veils it, veiling also the processes which result in what is called "genius," lies outside our limits, and it suffices to say that when a certain point is reached in social evolution, the old conditions reassert their power, and the truce to the struggle ends.

Perhaps enough material has been collected together to set the Ancient Stone Age men before the mind's eye as gathered into wandering tribes dependent for food on the chase: camping-out by the river-side under trees, or dwelling in huts built of branches, and resorting, as need arose or vicinity permitted, to the protection of cavern and rock-shelter. With the barbed spears and arrows they caught fish and shot fowl; with the more ponderous stone weapons they slew bigger game: mammoth, bison, rhinoceros, reindeer, and horse. The flesh, cut into pieces with flint knives, was cooked in vessels of wood or skin, into which were dropped hot stones as "pot-boilers." The bones were split for the marrow. The skins, scraped with flints, and sewn with bone needles threaded with sinew, covered the bodies against the often severe cold; even the hands, as portraits from the Pyrenees caves show, being protected with long gloves. The few and graphic touches in which Tacitus describes the Fennic (or Finnish) tribes may be applied to the earlier folk of drift and cave. "They are wonderfully savage, and miserably poor. Neither arms nor homes have they; their clothing is skins, their bed the earth. Their arrows, for want of iron, are tipped with bone. The women live by hunting, just like the men; for they accompany the men in their wanderings, and demand a share of the prey. And they have no other refuge for their little children against wild beasts or storms than to cover them up in a nest of interlacing boughs. Such are the homes of the young; such the resting-place of the old. Yet they count this greater happiness than groaning over field labour, toiling at building, and poising the fortunes of themselves and others between hope and fear. Heedless of men, heedless of gods, they have attained that hardest of results, the not needing so much as a wish," or, as it may also be translated, they "are beyond the need of prayer."

The scarcity of human bones in the Ancient Stone Age is of minor importance in presence of the proofs of man's tenancy of the globe during an enormous period, and at a low stage—indeed, the lowest stage—of culture. For the tools and weapons of drift and cavern are products of human skill; they have defined purposeful shapes; they evidence selection on the part of their makers, since they cannot be fashioned from every kind of flint. They are found, in striking correspondence of form, wherever man is known, or may be presumed, to have wandered over the earth, the extremes of the northern hemisphere excepted—in the alluvials of the East, the laterite or brick-earth of Madras; the river-gravels of sacred and classic lands; by the Sea of Galilee, and along the valley of the Tiber—in brief, in wellnigh every explored part of the world "from China to Peru."

CHAPTER IV.

THE NEWER STONE AGE.

ALTHOUGH this division is retained for convenience, it is more than probable that no hard and fast line can be drawn between the two Stone Ages. Stress is laid in most treatises on the subject upon the immense interval which separates the periods. The evidence of this is based on the different conditions, as the changes of climate; the altered distribution of land and water; the disappearance of old species of plants and animals, and the appearance of new species.

In the Neolithic Age Great Britain and Ireland no longer formed part of the continent. The valleys that had united those islands and the mainland had become submerged, a change

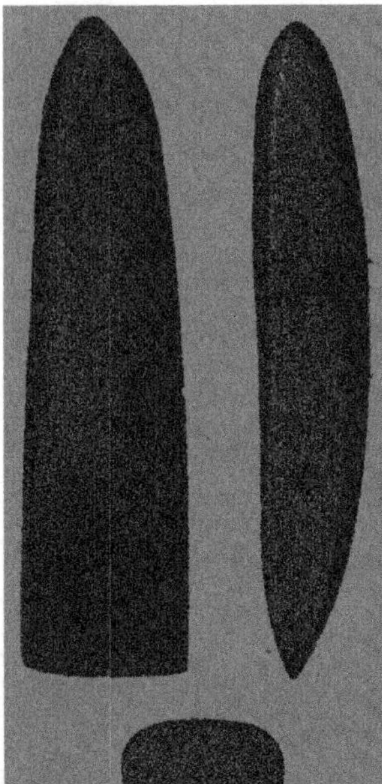

FIG. 28.—POLISHED OVAL CELT, WHITWELL, YORKSHIRE. (*Evans.*)

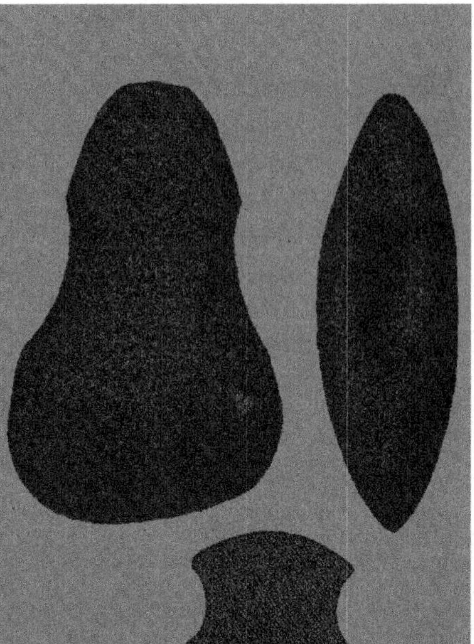

FIG. 29.—POLISHED CELT (WITH CAVITY ON EACH SIDE FOR THE FINGER), DUGGLEBY, YORKSHIRE. (*Evans.*)

which only a vast lapse of time brought about. The outlines of the map of Europe presented nearly the same features as at present. The area of the Mediterranean had sunk, separating Europe from Africa, the higher ground remaining as islands which are left like fragments of a sunken bridge. The big mammals, as the woolly rhinoceros and mammoth, were extinct; others had retreated to more northern and southern latitudes; the musk sheep to arctic zones; the lion, hippopotamus, and lynx to tropical zones. The animals found associated with Neolithic man represented —

FIG. 30.—POLISHED CELT, COTON, CAMBRIDGE. (*Evans.*)

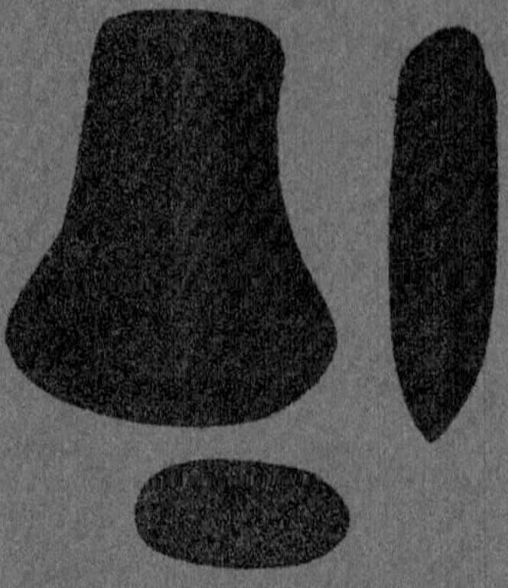

FIG. 31.—CELT FROM GRAVEL-PIT, NR. MALTON, YORKSHIRE. (*Evans.*)

some survivals, as the Irish elk, wild ox, wild boar excepted—species familiar to us. While the stone relics of Palæolithic times are found *underground*, in ancient river-gravels and in "caves and dens of the earth"; those of Neolithic times are *above ground,* or at slight depth; in surface remains, cave floors, camps,

FIG. 33.—POLISHED CELT AND ORIGINAL HANDLE, CUMBERLAND. (*Evans.*)

FIG. 32.—POLISHED CELT, GUERNSEY. (*Evans.*)

rubbish heaps, pile-dwellings, tumuli, and other burial-places. And while long and intermittent breaks appear to disturb the sequence of man's presence, at least in Europe, beyond the period of the cave-dwellers, his history, from the unknown time

FIG. 34.—PERFORATED AXE, HUNMANBY, YORKSHIRE. (*Evans.*)

FIG. 35.—AXE IN STAG'S-HORN SOCKET, CONCISE, L OF NEUCHATEL. (*Evans.*)

FIG. 36.—AXE-HEAD, POTTER BROMPTON WOLD. (*Evans.*)

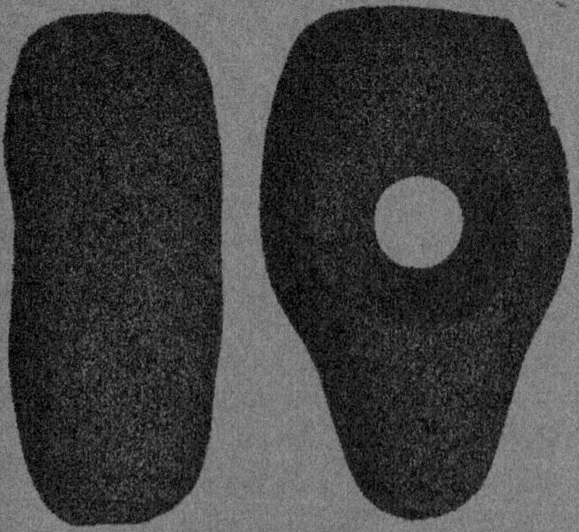

FIG. 37.—PERFORATED HAMMER, SCARBOROUGH. (*Evans.*)

FIG. 38.—HAMMER-STONE, HELMSLEY, YORKSHIRE. (*Evans.*)

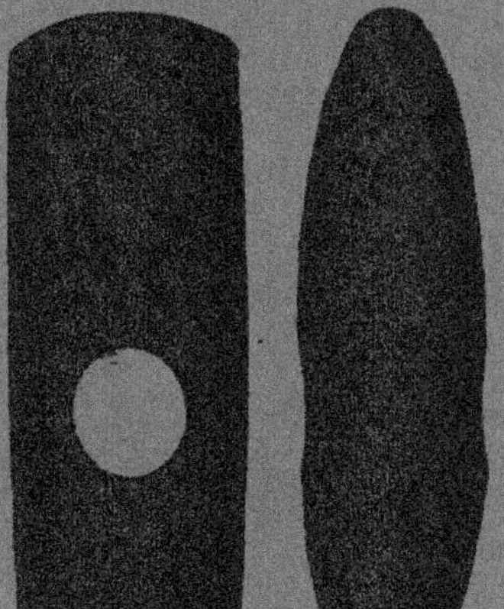

FIG. 39.—PERFORATED HAMMER, SHETLAND. (*Evans.*)

of the appearance of the earliest Neolithic people, is continuous to the present day. But, notwithstanding the enormous gap caused by the period during which the subsidence of land beneath the sea was going on, there is evidence which points to a continuous occupation of the British Isles, and of Europe and Asia, by the same race who gradually advanced in stages of culture, and who adopted the civilisation of somewhat higher races as this reached them by peaceful intercourse.

So that, weighing one thing with another, the balance tilts in favour of fusion between Palæolithic and Neolithic Ages. Not only are there abundant types of tools and weapons that are intermediate in character; but the oldest forms of

FIG. 40.—SCRAPER, RUDSTONE, YORKSHIRE. (*Evans.*)

copper and bronze implements are modelled on the patterns of the earlier stone and bone implements. And although changes brought

FIG. 41.—NARROW ADZE, OR PICK, BURWELL, CAMBRIDGE. (*Evans.*)

FIG. 42.—FLAKE SAW, WILLERBY WOLD, YORKSHIRE. (*Evans.*)

about the extinction or migration of the older fauna, those of the newer period may be survivors of animals who were contemporaries of

FIG. 43.—BORER, YORKSHIRE WOLDS. (*Evans.*)

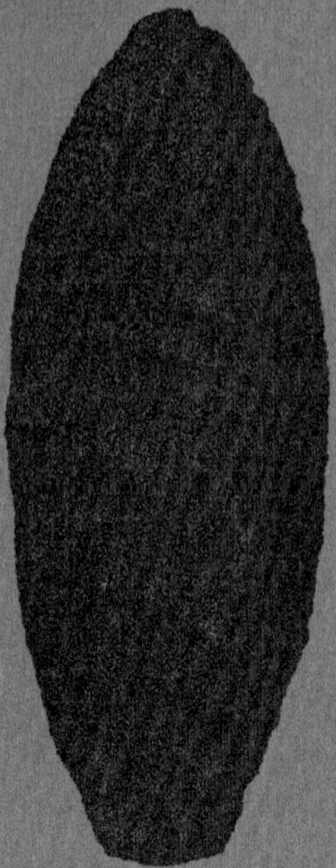

FIG. 44.—KNIFE, FORD, NORTHUMBERLAND. (*Evans.*)

FIG. 45.—CURVED KNIFE, FIMBER, YORKSHIRE.

the chipped flint workers of the Somme valley. The remains found in the cave of Duruthy show transition between the ages of ground and unground implements, as do like finds in the river-gravels of Sussex; and discoveries of a corresponding kind may be expected as researches are carried on.

FIG. 46.—LANCE HEAD, SNAINTON MOOR, YORKSHIRE (*Evans.*)

The supporters of the theory of continuity have suggested the name "Miolithic" or "Mesolithic" for the connecting period.

I. *General character of Newer Stone Age Implements.* As observed above, Neolithic implements do not occur in deep-lying or sealed-up deposits like stalagmitic beds, but either on the surface or very near it. Where the soil has been used only for tillage or pasture, it has been but superficially disturbed; where it is rocky and barren it has often not been disturbed at all. Consequently, an enormous number of implements either meet the eye or are turned up by plough and harrow, or uncovered by

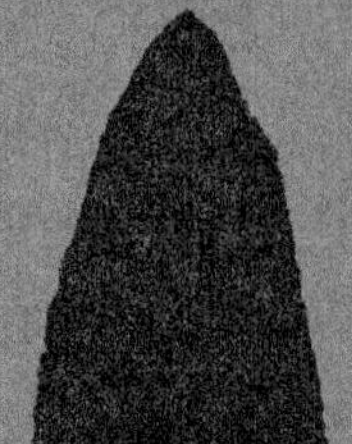

FIG. 47.—KNIFE, SAFFRON WALDEN (*Evans.*)

the action of rain. As recollection of a Stone Age died away, they have been looked upon with

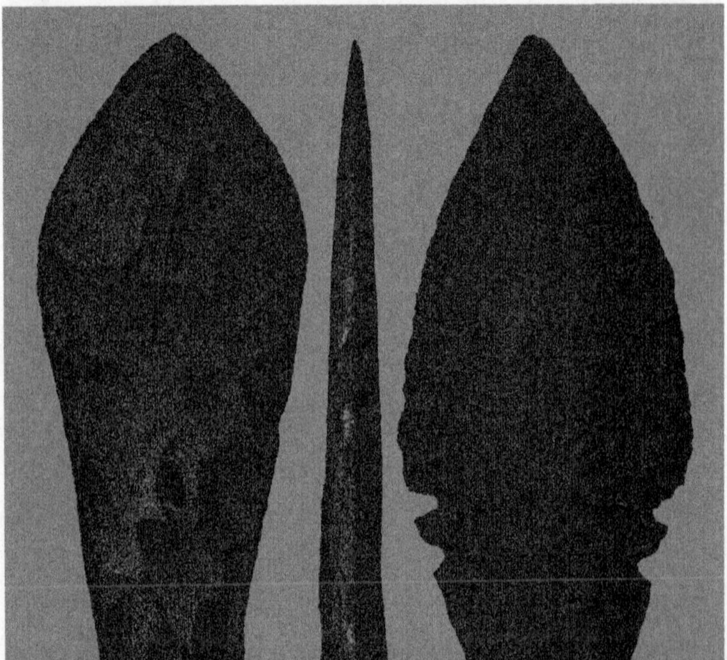

FIG. 48.—DAGGER, THAMES. (*Evans.*)

FIG. 49.—NOTCHED SPEAR-HEAD, BURNT FEN, ELY, (*Evans.*)

veneration, and have given support to a mass of crude ideas and superstitions, about which more presently. The most common form of Neolithic implements is that known as the celt, probably

FIG. 50.—INDIAN AXE FROM THE RIO FRIO, TEXAS.

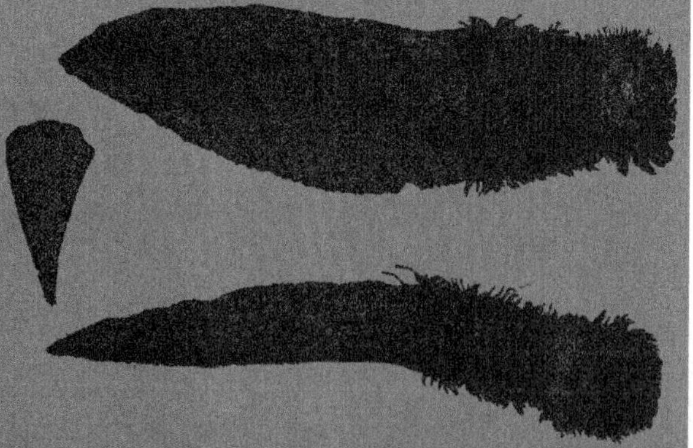

FIG. 51.—FLINT KNIFE, AUSTRALIA. (*Evans.*)

so called from the Latin *celtis*, or *celtes*, a chisel. The shape of this instrument is generally that of a flat blade, approaching an oval in section, with the sides more or less straight, and with one end broader and also sharper than the other. The length varies from two to sixteen inches, and the stone of which celts are made varies according to the kind most accessible. A representative set of Neolithic implements would comprise, in addition to the celts, stone tools allied to picks, small hand-chisels and gauges, perforated axes, some sharp at the end, others shaped like adzes, saws, hammers and hammer-stones, grinding stones, querns, sink-stones for nets, whetstones, scrapers, borers, awls, drills, and knives. The purposes to which these would be applied are as numerous as the needs of man. Modern savages use like tools for cutting

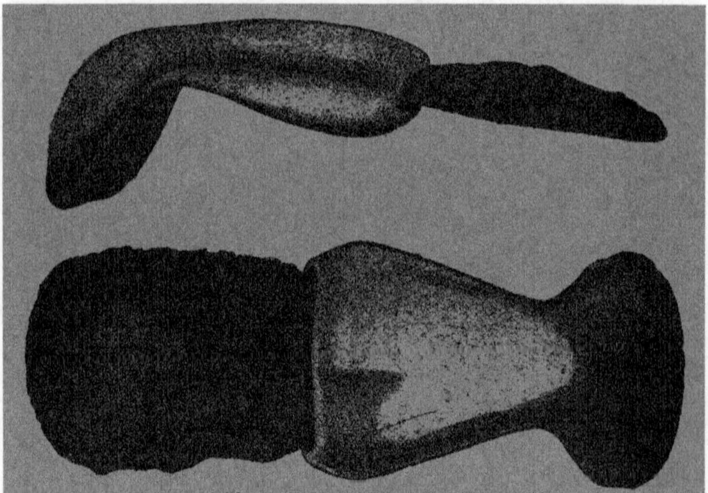

FIG. 52.—ESQUIMAUX CRAPER. (*Evans.*)

timber, scooping out canoes, dressing posts for huts, grubbing up roots, killing animals, and scraping the flesh from their bones. Then there are the implements required for domestic purposes, while for war and the chase there were daggers, javelin-heads, sling-stones, bolts, lance and arrow-heads, some of these last of exceeding beauty and finish. Bone lance-heads, pins and needles, were also used, and staghorn was made into hammers and axes. The women's spindle-whorls were of stone, and their personal ornaments, the simplest form of which was the button or stud, were of jet, shale, and amber. The antiquities thus briefly summarised occur in the upper layers of cave-deposits, in peat bogs, coast-finds, refuse-heaps, and pile-dwellings; in tumuli, barrows, and various stone structures, as cromlechs, dolmens, &c., scattered over the world.

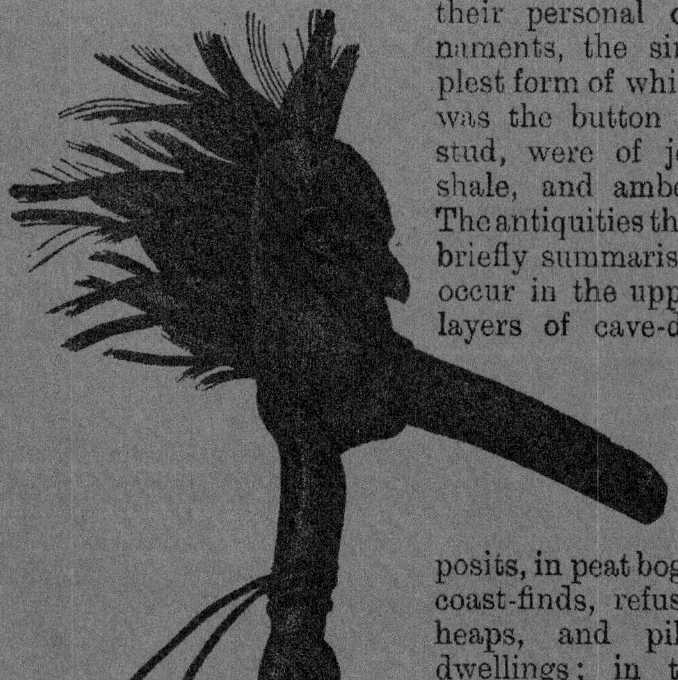

FIG. 53.—WAR AXE, NOOTKA SOUND. (*Evans.*)

The superstitions just referred to have gathered round celts, which are known among rustics as "thunderbolts" or "thunder axes"; round arrow-heads, or "elf-shot"; and round spindle-whorls, called "fairy-millstones" and "pixy's grindstones" by British peasantry. For

FIG. 54.—JAVELIN, BARROW, NEAR STONEHENGE. (*Evans.*)

FIG. 55.—ARROW-HEAD YORKSHIRE WOLDS.

FIG. 56.—ARROW-HEAD, BRIDLINGTON. (*Evans.*)

ages it was a belief shared by the learned and un

FIG. 57.—STEMMED ARROW-HEAD, YORKSHIRE WOLDS.

FIG. 58.—STEMMED ARROW-HEAD, YORKSHIRE WOLDS. (*Evans.*)

FIG. 59.—STEMMED ARROW-HEAD, IWERNE MINSTER, DORSETSHIRE.

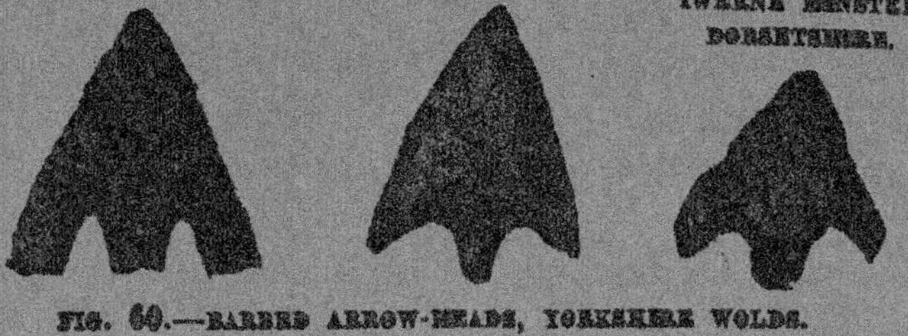

FIG. 60.—BARBED ARROW-HEADS, YORKSHIRE WOLDS. (*Evans.*)

learned that with the flash of lightning there fell a solid body, which is called the thunderbolt or

FIG. 61.—SPINDLE WHORL, SCAMPSTON, YORKSHIRE.

FIG. 62.—SPINDLE WHORL, HOLYHEAD. (*Evans.*)

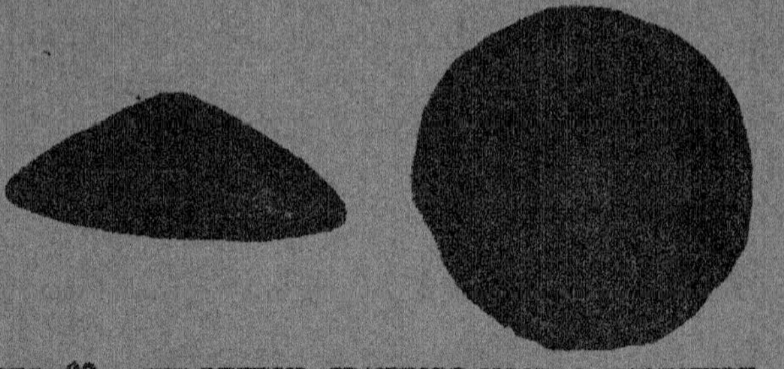

FIG. 63.—JET BUTTON, CRAWFORD MOOR, LANARKSHIRE. (*Evans.*)

thunderstone, as expressed in the dirge in "Cymbeline"—

> "Fear no more the lightning flash
> Nor the all-dreaded thunderstone;

and it is these Neolithic relics—axes and arrow-heads—to which celestial origin has been assigned. They were known to both Greeks and Romans, as they are to the Indians of Nicaragua, as thunderbolts; the Germans and Scandinavians called them Thor's hammers, and both among them and other European peoples they were credited with miraculous powers in healing the sick and warding off the dire effects of the evil eye. The natives of the Gold Coast,

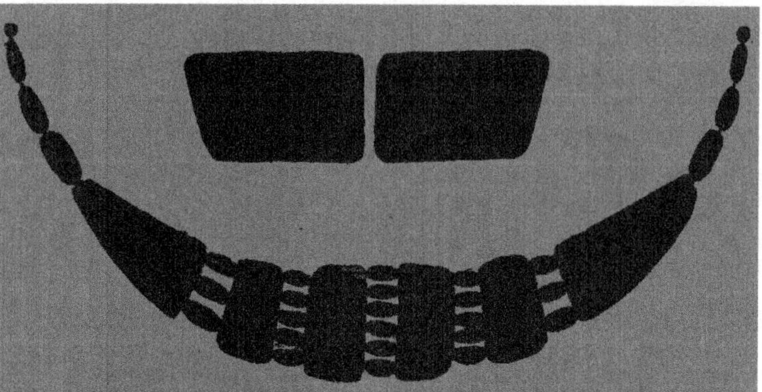

FIG. 64.—NECKLACE OF JET, STUDDED WITH MINUTE SPOTS OF GOLD. FOUND IN URN WITHIN BARROW, ASSYNT, ROSS-SHIRE. (*Evans.*)

when finding them on the ground after heavy rains have washed them out of the soil, use them as medicine by scraping the dust from them into water, and laying them in places sacred to the gods. In Brittany the travelling umbrella-mender asks on his rounds for *pierres de tonnerre*, and takes them in payment for repairs. In India they are valued as charms whose possession brings good luck to their owner, and whose loss is the signal of his ill-fortune

The arrow-heads are also called elf-shot and elf-stones by the country folk of Britain and Ireland, in out-of-the-way places, it being their belief that these flint weapons were shot by the elves, or fairies, at men and cattle. Thus Robert Gordon, of Straloch, an accomplished country gentleman of the north of Scotland,

FIG. 65—ELF-SHOT. (Worn by an old Scottish lady for half a century.)

writing in 1654, tells how one of his friends, travelling on horseback, found an elf-bolt in the top of his boot, and how a gentlewoman of his acquaintance, when out riding, found one in the breast of her habit These "elf-bolts" were often mounted in silver, and worn as charms against poison and witchcraft; sometimes they had a hole drilled through them, so that they might be dipped in water and endow it with healing virtues. It is a popular belief in the Highlands that when cattle are sick they have been struck by an elf-dart, and the quack doctor, feeling the animal all over, produces the magic weapon as if found in the skin, as the "medicine man" among barbaric folk, in curing the toothache, always extracts a worm to the satisfaction of the patient. Then the elf-dart is put into water, which is given to the animal, and, of course, a cure effected. Sir John Evans quotes from Pitcairn's "Criminal Trials" the description of a cavern where the arch-fiend carries on the manufacture of elf-arrows with the help of his attendant imps, who rough-hewed them for him to finish; and also cites from Wilson's "Pre-Historic Annals of Scotland" a letter from one Dr. Hickes to the famous diarist, Pepys, "recording that my Lord Tarbut, or some other lord, did produce one of these elf-darts, which one of his tenants or neighbours took out of the heart of one of his cattle that had died of an usual death" (*sic*). Dr. Hickes had another strange story, but very well attested (!), "of an elf-arrow that was shot at a venerable Irish bishop by an evil spirit, in a terrible noise louder than thunder, which shaked the house where the bishop was." Bosnian peasants wear them in necklaces as charms, and the long range of this superstition is seen in the flint arrow-heads sometimes forming the central pendant in necklaces found in Etruscan tombs, and, among the Greeks and Romans, adorning the diadems of their gods. The Italian peasants keep flint arrow-heads to preserve their houses from lightning, and in some instances carry them about their persons as amulets. The miniature arrow-heads of cornelian worn by the Arabs of North Africa and other people as charms, or as good for the blood, have assumed the shape of a heart, and thus become symbols of love. So abundant and world-wide are the examples from which this haphazard selection is made, that the rest of the book might be filled with them. But it will be seen presently that they are nearly related to that adoration of stones as objects of worship which has been a prominent feature of barbaric religion.

II. *Remains found in Coast-finds and Shell Mounds.* Turning now to the relics of earliest Neolithic, or, perhaps, Mesolithic deposits, these are found in accumulations of rudely worked flints lying near the sea-shore in various parts of the world. They are known as coast-finds, and, judging from the character of the chips, were probably workshops. For the manufacture of implements in selected spots must have become a considerable industry, as men multiplied and, perhaps, divided the labour of chipping flints from the labour of using them.

But of more importance than these are the refuse-heaps known as kjökken-möddings, or "kitchen-middings," once mistaken for natural formations in the shape of raised beaches, but since discovered to be the sites of ancient fishing and hunting settlements. They are mounds of various size, sometimes one thousand feet long, and two to three hundred feet wide, and are composed of castaway refuse—myriads of oyster and other shells; bones of the stag, roedeer, dog, and other animals still extant; of wild duck, wild swan, caper-cailzie, and other familiar birds; of cod, herring, flounder, and other deep-sea fish. All the marrow-yielding bones had been split open. Besides these, the surer tokens of man's presence were found in implements of stone, bone, and wood, and—what marks the chief distinction between Palæolithic and Neolithic remains—some rude pottery. Some charred wood and burnt substance of a sea plant, which perhaps yielded salt, were found, but there were no traces of any grain. The stone implements comprised axes, flakes, hammers, awls, lance-heads, sling-stones—generally of rude type. The absence of bones of any other domesticated animal makes the remains of the dog—without doubt, the earliest animal tamed by man—highly interesting as marking the approach to a settled kind of life. Professor Steenstrup proved this domestication of the dog, the conversion of "the brother of the wolf into the guardian of the flock," by an ingenious experiment. He noticed that nearly all the long bones of the animals taken from the mounds were reduced to their shaft, and that the heads were irregularly broken. He then experimented with dogs, and found that they gnawed the heads of the long bones, and rejected the hard and solid shafts. The bones of deep-sea fish show that these rude colonies of fishers and hunters, had probably, ventured on the ocean; perchance sons of

"the first that ever burst into that silent sea";

in "dug-outs" furnished with nets of twisted bark, or of bast, or some other fibrous plant.

The larger number of kitchen-middings in Denmark follow the present or former coast-lines; but some of them are inland, evidencing the advance of the land on the sea since their abandonment, which involves their very high antiquity. Moreover, the oyster, save in stunted form, has practically disappeared from the Baltic waters, owing to their having changed from salt to brackish. This shows that Denmark was formerly more intersected with fjords or narrow seas; at any rate, that there was freer communication between the Baltic and the Atlantic ocean. Similar refuse-heaps occur in various parts of the world, on the banks of the great American rivers; on the seaboard of South Africa, South America, Australia, and wherever man has eaten fish and left their bones behind him in his primitive migrations. These were always along coast-lines and by river banks.

III. *Races of the Newer Stone Age.* Crossing from the Continent to Britain, what manner of men do we find there in early Neolithic times? It may be assumed that the blank which divides cave man at his highest from Neolithic man at his lowest exists only because the materials that could have filled it are vanished, never to be recovered. That blank marks the interval during which Britain "arose from out the azure main," and whether or not the rude folk living there, thus separated from Europe, were the descendants of the cave-men, or whether they had arrived before the dividing waters filled the valley of the

North Sea or widened the Straits of Dover, we cannot say. Certain it is that man was then in course of advance from the hunting to the pastoral stage.

Britain was then, and till long within the historic period, a land of swamp and forest, of chill summers, and drenching climate. The inhabitants lived in caves and rock-shelters where these were to be had, otherwise in huts of boughs or loam. When the dead were buried, they were laid in caves, or in tumuli or long "barrows" (Anglo-Saxon *berg*, a hill or hillock; the term "cairn" used in Scotland is from Gaelic *carn*, "a heap"), one evidence of the later date of the barrows being in the fewer remains of wild animals found in them. The human skulls also belong to the older or "dolicho-cephalic," that is, "long-headed" type. It should be explained that skulls are measured by the relation of breadth to length, that is, from back to front, taking that at one hundred. If the breadth is under eighty the skull is called "long-headed," if it exceeds eighty, it is called "brachy-cephalic," or broad-headed. There are other sub-indices, but with these we are not concerned. The shape of the skull, varying, as this does, so little in races, is, on the whole, the best test that can be applied to their identification. No traces of metal are found in any of the long barrows, and pottery is rare. Some of these graves are as much as four hundred feet long, and fifty feet wide. They are modelled upon the natural caves which were the abode of the living; for among all barbaric people the home of the dead has been a copy of his dwelling when alive, a custom the meaning of which will appear later on. These "long-headed" Neolithic folk, who are generally known as "Iberian" (they are also variously called Berbers, Basques, Silurians, and Euskarians), were small-limbed, swarthy-complexioned, and with dark hair and eyes. Their representatives are widely distributed nowadays, and in these islands are found chiefly in the west of Ireland and some parts of Wales and the Highlands. A typical example is thus described by the late Mr. Campbell, of Islay, as met with in his travels when collecting West Highland folk-tales from the mouths of the people. "Behind the fire sat a girl with one of those strange foreign faces which are occasionally to be seen in the Western Isles, a face which reminded me of the Nineveh sculptures, and of faces seen in St. Sebastian. Her hair was as black as night, and her clear dark eyes glittered through the peat-smoke. Her complexion was dark, and her features so unlike those who sat about her, that I asked if she were a native of the island, and learned that she was a Highland girl." The persistence of a race of which specimens meet us everywhere, to which, in high probability, some of the readers of these pages belong, and which, as there is every reason to think, is slowly reasserting its old supremacy, is thus shown by Boyd Dawkins.

"Through all the troubles which followed the conquest of Gaul by Cæsar, and of Britain by Claudius; through all the terrible events which accompanied the downfall of the Roman Empire, causing the Britons to be exterminated over a large part of England, and the almost total extinction of the ancient type of Roman in Italy, the Iberian lived, and still is found in his ancient seats, with physique scarcely altered, and offering a strong contrast to the fair-haired Celtic, Belgic, and German invaders. The Iberian race is known to the ethnologist and historian merely in fragments, sundered from each other by many invasions and settlements of the Aryan race. It is shown by the researches into caves and tombs to have been in possession of the whole of Europe north and west of the Rhine in the Neolithic Age, and has been traced by Dr. Virchow into Germany and Denmark."

Towards the close of that Age another race appears in Britain. They contrast in every way with the Iberian, being tall, large-limbed, broad-skulled, with fair hair and blue eyes. They are known as "Celts," one of the three races—the others being the Aquitanians and Belgæ—who were scattered over North-Western Europe in the time of Cæsar.

These "ancient Britons," whose tall stature so impressed the Romans, lived in "hut circles," or "pit dwellings," and buried their dead in round barrows—copies of their houses. The log or wattle huts have perished, only the traces of their ramparts, crowning many a hill, remaining; but of the subterranean dwellings abundant specimens survive. These are from seven to ten feet in depth, were roofed over with boughs or flagstones, and entered by a sloping shaft, or a narrow "man-hole" doorway, near the top.

Fragments of pottery, ornamented, after the fashion of primitive ware, by nail or finger-marks, or by impressions of a cord twisted round the soft clay; "mealy-stones," or querns of a type barely extinct in Scotland yet; bone needles, spindle-whorls, roughly-chipped celts and arrow-heads, but—at least among the earliest deposits—no traces of metals, connect them with a people who spun their clothing, planted corn, and, as the bones of oxen, horses, and pigs show, had tamed some of the wild animals. Of course the caves were artificial, serving as places of shelter and concealment, so that, as Tacitus says of like dwellings among the Germans, "should an enemy approach, he lays waste the open country, while what is hidden and buried is either not known to exist, or else escapes him from the very fact that it has to be searched for." The dwellings of primitive people, like the implements with which they make shift, are of a common type, or, at least, are reducible to two or three classes, and the underground earth-houses of Europe have their counterpart in Asia, Africa, and America. Speaking of some native tribes in Central Africa, Mr. Stanley describes them as living "in deep pits with small circular mouths leading to roomy apartments," and the same places of shelter and refuge may be traced across the world. Mr. A. H. Savage Landor describes the remains of pit-dwellings, which abound all over Yezo and the Kurile Islands beyond, the implements in them being ground adzes, arrow-heads, bone and bamboo arrow-points, mingled with fragments of pottery, and bones of foxes and other animals. To the raisers of the stone beehive-shaped huts and such like dwellings may be traced the great mass of barrows and kindred structures that in a large degree give us the desired key to the thought of pre-historic man before it was embodied in oral tradition and written on monument, papyrus, or wax.

IV. *Earth and Stone Monuments of the Neolithic Age.* Death, which sweeps us all away, has preserved the records of life as nothing else has done. The surface of the earth, wherever man has spread himself, is studded with his sepulchres. They range from the survival of the ancient mound in the little heaped-up grave of Christian churchyards to the huge cairns, called pyramids, in which the Pharaohs were laid, and the stately monument which Queen Artemisia raised over the body of Mausolus, whence the name *mausoleum.* Before writing in its earliest form—"picture-writing"—was invented, a few loose stones, or a single large stone, or, where these were lacking, a heap of earth, would mark without inscription the spot where the dead were laid, and out of these primitive monuments have been developed huge tumuli and stone sepulchres. Not that every mound is a burial-place, some being boundaries, or defensive earthworks, or moot or meeting-places, like the great mound known as Silbury Hill in Wiltshire, in which no skeletons have been found.

Barrows are either long, or more or less circular. They are chambered or unchambered, that is, either raised over a vault of upright stones, or simple excavations in the mound. The use of the same barrows as burial-places has not been limited to any one age or race, but the earliest interments are easily distinguished from the latest, not only by their position, but by the articles which, in a large number of cases, are

FIG. 66.—FROM TUMULUS, AT LE TUS, GUERNSEY.

found with the skeletons. Some of the more important tumuli are as much as four hundred feet long, and, in certain instances, are approached by an underground gallery leading to the sepulchral chamber, resembling the passage to the "yurts" of the Siberians and the "gamme" or underground dwellings of the Lapps. The primary interments in the long barrows are of the long-headed Iberians exclusively: the bodies being buried either at full length or in a crouched or contracted posture; resting on the haunches, as was the custom among the ancient Peruvians, and as is the custom among the Andaman Islanders. This probably represented the squatting position of the deceased during life. Everything, it must be remembered, was done to make the grave a copy of the house of the living, and to surround the corpse with objects familiar to it in life—in short, to make him feel "at home," and thus "lay the ghost." So the long barrow is modelled on the long-galleried cave.

FIG. 67—PLAN OF TUMULUS, GUERNSEY, SHOWING POSITION OF SEPULCHRAL CHAMBER.

Speaking of the primary interments in them, the round or oval barrows are the burial-places of the broad-headed "Celts" exclusively. They were so used until the close of the Neolithic Age, when the practice of burial was superseded by that of burning, the ashes being preserved in a stone cist or urn. These round barrows are modelled on the hut-circles or pit dwellings, and the objects found in them are similar in character so those yielded by the dwellings. Celts, flakes, arrow-heads, and pottery lie jumbled together, many of the articles having been purposely broken so that their spirits might be freed to join the dead owner and serve him, as the things themselves had done during his life.

V. *Primitive ideas about Spirits and an After life.* At this point we may pause to ask what light the contents of his burial-places throw upon man's ideas about the spirit-world, since, these being known, we hold the key to many a strange rite and custom. The answer can be only crudely outlined here. In drawing the brief sketch of primitive man's explanation of his surroundings, the chief feature noted was his confusion of life and motion, which led him by easy stages to belief in spirits everywhere and in everything. This belief was confirmed by the events of daily life, if those events are not, as some think, the prime cause of the belief itself.

Children and animals lie curled up during sleep; only adult man, it is said, has contracted the habit of sometimes lying on his back. Although dreams will happen in whatever position we lie down, we know from experience that the more disagreeable ones happen when we rest on our backs, and if, when doing so, the stomach is overloaded with food or refuses to digest it, we know that bad dreams of the "nightmare" kind result. And, in the case of barbaric folk, with whom the supply of food is often fitful, the heavy gorges in which they are prone to indulge make their dreams often uncanny. But whether these be pleasant or the reverse, there come into them apparitions of the dead and living, with whom the sleeper talks or feasts or fights, whom he joins in the chase or war-dance—in short, living the old life, the waking life, over again in all its reality, with much of strangeness added by the odd elements of the dream.

All which is clear enough to us, however much our dreams, especially the gruesome ones, haunt our waking moments, because we have learned what vagaries of the brain arise through partial sleep. But by the barbaric mind the events dreamed of are believed to have really happened; and therein lies not only the vast difference between our ways of looking at the thing, but the key to the explanation of the beliefs of all the lower races, and to such survivals of those beliefs as exist — in no small degree — among higher races.

Everything being thus regarded as real, the dead who appear in dreams are alive, and return to their old haunts, or receive visits from their friends or foes in some place of which the sleeper dreams. His squaw, lying wakeful by his side all night, may tell him, when she hears what he has done, that he has not left his bed. But he knows better than that, although she may speak truth; and if his body did not move, what did move?

Daily experience helps him to an answer. Other people are seen sleeping, moving restlessly, and then after a time waking-up. Or they are stunned by a blow, or falling down in a fit or swoon, lie helpless or speechless, perhaps for days, and then recover. Or they scream and toss with pain, and in other ways lie stricken, and then fall into a long, long sleep, and never wake again. Or in some more direct and violent way the same result happens: they die. To what other conclusion can the barbaric mind come than that everybody has *another self*, as it has been called, which does the things dreamed of, which leaves a man for a time when he is asleep or in a fit, which leaves him altogether when he dies, but comes back, and seems the very man himself to the dreamer?

The Malays do not like to wake a sleeper, lest they should hurt him by disturbing his body while his soul is out. When the Greenlander dreams of hunting or fishing or courting, he believes that the soul quits the body. The Melanesians say that the soul "goes out of the body in some dreams, and if for some reason it does not come back the man is found dead in the morning; when a man faints, his soul really starts on the way to Panoi (the underworld), but is sent back; the other ghosts hustle him away from the mouth (of Panoi), or his father or friend turns him back, telling him that his time is not yet come." And in the Solomon Islands if a child starts in its sleep it is believed that some ghost is trying to snatch away its soul, as also, if it sneezes, the notion is that "a ghost is drawing the soul away." But the superstitions which connect sneezing with spirits would fill a long chapter. To the Indians of Guiana dream-acts and waking-acts differ only in this—that the one are done only by the spirit; the other by the spirit in its body. The following example, given, among others, by Mr. im Thurn, must suffice.

"The morning when it was important to me to get away from a camp on the Essequibo River, at which I had been detained for some days by the illness of some of my Indian companions, I found one of the invalids so enraged against me that he refused to stir. For he declared that I had taken him out during the night and had made him haul the canoe up a series of difficult cataracts. Nothing could persuade him that this was but a dream, and it was some time before he was so far pacified as to throw himself sulkily into the bottom of the canoe. At that time we were all suffering from scarcity of food, and, hunger having its usual effect in producing vivid dreams, similar events frequently occurred. More than once the men declared in the morning that some absent man, whom they named, had come during the night, and had beaten or otherwise maltreated them and they insisted on much rubbing of the bruised parts of their bodies."

What is this "other self" like? It is seen and heard, yet never handled. Looking into water a man observes a figure of himself, or when the light is present, a dark copy of himself, longer, or shorter, or aslant, before, or behind him. Both shadow and reflection mimic his movements. Again: Wherein seems the difference between a dead and a living person? Till decay sets in there is little outward change, except that the man moves not;

> "Though one should smite him on the cheek,
> Or on the mouth, he will not speak;"

and no warm air is felt when his lips are touched.

So the "other self" comes to be conceived of as a sort of vapour, compounded of breath and shadow and reflection; "a certain soul and semblance, though substance there be none," as Achilles says when he clasps the shade of Patroclus in Hades. And in every language, from that of the barbaric Ainu to the classic and our own, the word for "spirit" and for "breath" is the same. Here civilised and savage meet on common ground, the answer of each has the same vagueness to the question, what is the soul like? And when we seek to envisage its "questionable shape" it is conceived of as a sort of replica of the body all the world over, throughout races at every level of culture. That Christian art could accomplish nothing higher is seen in the frescoes on the walls of the Campo Santo at Pisa, where the soul is portrayed as a sexless child emerging from the mouth of a corpse; in an elaborately sculptured monument over the tomb of Bishop Giles de Bridport in the east transept of Salisbury Cathedral, where the soul is represented as a naked figure being carried by an angel to heaven;' and in every other expression of the sculptor's and painter's art.

And this "other self," whether its destiny be the upper or nether world, hovers about its old home and haunts. When whaling vessels first touched at Banks' Islands, the natives argued that the crew were not men, because, if they were, they would be black. They must therefore be the ghosts of men who had lived. When Bishop Patteson landed at Mota, one of the same island group, it was held that he was a ghost, and on his retiring from the heat and crowd to an empty house, the owner of which had lately died, the natives were satisfied that he was the ghost of the dead man, for did he not know his home? The grave, however, as containing the body, was the more direct place of communication with the dead, and therefore with the spirit. Round it gathered from primitive times customs which, in one form or another, are in full force to this day. The belief that the dead man still required the necessaries of life made the supply of these a solemn duty, and while piety and sympathy aided the discharge of this, prudence prompted care that no neglect should offend those who perchance had acquired new power to help or harm the living. Hence, as shown by the food and libations brought to every grave, barbarian, classic, and Christian, and by the articles deposited with the corpse, each of which was, in the lower culture, also believed to have its "other self," nothing was left undone to make the dead content with their new lot. From the Reindeer period to the Christian era it is the same. The child's rude ivory doll found in the Dordogne cave has its counterpart in the terracotta dolls and animals found in the tombs of Pompeian children, and in like objects in the catacombs of Rome. Perhaps one of the most touching relics, linking past and present, is that of the skeleton of a young woman clasping a child found in a round barrow on Dunstable Downs, the grave being edged with fossil, *echini* or sea urchins, which the peasants call "fairy loaves."

In examining British tumuli Canon Greenwell frequently found holes below the natural surface, inside the barrow, usually only filled with the soil, but occasionally with human bones, fragments of pottery and charcoal. The probability is that these cup-like hollows were receptacles for food or drink for the use of the dead, like the cups filled with holy water sometimes seen on Roman Catholic graves on the Continent, the use of which, although now symbolical, is a survival of primitive custom. That custom has abundant illustration among modern barbaric folk. To cite an example or two, the Bodo of India carry the dead man's share of food and drink to the grave, saying, "Take and eat, heretofore you have eaten and drunk with us, you can do so no more; you were one of us, you can be so no longer; we come no more to you, come you not to us." Then each man throws on the grave a thread bracelet, which he breaks off his wrist in token that the tie between dead and living is snapped, and the company afterwards bathe before feasting. This idea of impurity through contact with the dead is world-wide, as shown in the lustrations practised by every race. At their funeral feasts the Solomon Islanders throw a piece of food into the fire, saying to the dead "This is for you," and even at their daily meal, when the oven is opened, a portion is put aside for the dead. The Celestial "keeps his coffined parent for years, and serves him with meals as if he were still alive." But China is the home of ancestor-worship on the hugest scale; not only are the dead treated as if alive, but as also capable of being promoted, by Imperial decree, to higher rank, as are the gods themselves. Curious examples of this appear from time to time in the *Pekin Gazette*, which is the oldest newspaper in the world. For example, when the dragon-spirit of Han Tan Hien added to the favours which it had already conferred by visiting certain provinces with much-needed rain, it was ordered that the god should be invested with the additional title of "the Dragon-Spirit of the Sacred Well." In another *Gazette* the "Director-General of Grain Transports prays that a distinction be granted to the god of winds, who protected the dykes of the Grand Canal; whereupon the Board of Rites is called upon for a report." When Kwangte, a great commander, who was deified eight hundred years after his death as the God of War, was believed to have aided the Imperial troops in crushing the Taiping rebellion in 1855, the Emperor issued a decree promoting him to equal rank with Confucius. The Roman Catholic Church does precisely the same thing in canonising the more distinguished defenders of the faith centuries after their death. Barbarian and civilised are alike in this, as in so much else; and thus it is that the Tonga belief as to the existence of caste in another world obtains among civilised people. The Hindu offers the funeral cakes, with which may be classed the "simnels" still made in England, and the "soul-bread" in Belgium. But, more often, special seasons are dedicated to the service of the dead, as is the Fiji offering of the earliest yams; in the Slavonic custom of laying food on the grave in springtime; in the Bulgarian feasts in the cemeteries on Palm Sunday, when the remains are left to be eaten by the dead; in the last day of the feast of Bacchus, the All Souls' Days of the Greeks; and in the commemorative festival of like name among Christians.

The primitive and grosser idea is that the spirits ate and drank of the actual things offered, as among the Congos, who send the meals through an opening in the grave to the mouth of the corpse. The Greek offerings of food and drink at the tombs made their mocking satirist, Lucian, ask if the libations filtered through the ground to Hades. But the source of the idea is clear enough. Solids, when burnt, pass away in the form of smoke, or, when left anywhere, lose their moisture and shrivel up; liquids, poured on the ground or heated, evaporate; hence the conclusion that both are consumed by the shadow-soul. For what does the barbaric mind know of the processes of heat and evaporation? Later on, the belief prevails that the incorporeal spirit consumes only the essence of things; the "sweet-smelling savour" satisfies the ethereal nostrils, and the coarser parts are left behind.

The bones often found near Neolithic barrows evidence the great antiquity of the custom of feasting at the grave as a commensal act between the dead and living, and the cloven skulls of human beings associated with these remains tell of the sacrifice of slaves and prisoners, that the departed, especially those of higher rank, may not lack attendants. Other human bones point to the self-sacrifice of wives that they might join their dead husbands. But the care and completeness with which, to the smallest detail, the supposed needs of the dead were supplied were largely due to the fear of his coming back. The unburied soul hovered restless and troublesome, denied admittance to the underworld. "Bury me with all speed, that I pass the gates of Hades," prayed the spirit of dead Patroclus to Achilles in a dream; and once housed in the tomb, the chief thing was to keep the body there. Earth or stones were heaped upon it to prevent its return to the living; the legs were tied to the body, as among the Australians; pins driven into the feet, as among the Icelanders; or a stake through the body, as among ourselves, in burying the suicide by the puzzling four cross-way. All sorts of dodges were used to baulk any attempt of the dead to return. In the Solomon Islands the funeral procession comes back by another road, lest the corpse should follow; the Hottentots remove the dead by a special opening in the hut; the Minahassees of Celebes, before burying the corpse, thrust it into a hole in the floor, and carry it three times round the house; in Homer the dead Patroclus was laid with the feet towards the door of the tent; and in some parts of Europe to this day the body is carried from the house with the feet outwards, or, as among the peasants of Branden-

FIG. 68.—HOLED DOLMEN, CIRCASSIA. (From Original Drawing on the spot by William Simpson, R. I.)

burg, when the coffin has left, a pail of water is emptied at the door to hinder the ghost from entering. For spirits and witches were supposed to be unable to cross water, which may perhaps account for the frequent burial of the dead, as among the Celts, in islands.

But, in these matters, the barbaric mind, as indeed that of higher folk, is a tangle of contradictions. For while, as the foregoing examples show, among every people pity and fear unite to secure rest and comfort to the departed, we find frequently in cromlechs, dolmens, funeral urns, and in tombs of the later Roman period, little openings the probable object of which was to give ingress and egress to the ghost. This is, in fact, the reason assigned by the Iroquois: that the soul may come out and go in at pleasure. Among the rustic folk of civilised countries the same idea takes the shape of opening a window in the room of the dying so that the spirit may depart unhindered; and is expressed in the German saying that one should not slam a door lest a soul gets pinched in it.

Connected with openings in tombs is the practice of the trepanning of skulls, or the boring of a hole in them. It was practised in pre-historic times, as shown by skulls found in Neolithic caves both in the Old World and the New. The circular pieces of bone thus cut away were often worn as charms; but their removal was probably due to the desire to afford the soul of the dead free passage, and to cure diseases of epileptic and other nervous kinds among the living. The trephine used in modern surgery for removing portions of bone in cases of injury to the head is thus the lineal descendant of the flint saw of the Neolithic operator. His ideas were at the level of those of the barbaric mind all the world over in its explanation of disease as due to the presence of evil spirits through sorcery or other black arts hence the making of a hole to let out the demon.

FIG. 69.—HOLED DOLMEN, CIRCASSIA. (From Original Drawing on the spot by William Simpson, R.I.)

Among the Melanesians, whose ideas as to the causes of sickness may be taken as largely representative of barbaric belief, it is always, except in common and slight complaints, said to be caused by a ghost or soul of a dead person, as contrasted with a spirit or supernatural being that never was in a body. "It happens, indeed, as in the New Hebrides, where spirits are the chief objects of religious regard, that a man knows that he has trespassed on a sacred place belonging to some spirit, or has an illwisher who has a spirit for a helper, and supposes therefore when he is ill that a spirit has brought his sickness on him. But generally it is to the ghosts of the dead that sickness is ascribed; recourse is had to them for aid in causing and removing sickness; and ghosts are believed to inflict sickness not only because some offence has been done against them, or because one familiar with them has sought their aid with sacrifice and spells, but because there is a certain malignity in the feeling of all ghosts towards the living, who offend them by being alive." For death, like disease, is not regarded as a natural event; hence, amongst reasons already cited, the appeasement of the unwilling departed. Language bears its testimony to barbaric theories, as in "seizure," "possession," being "out of one's mind"; and in the folk-lore of North England, in which "casting out the ague" was but another name for "casting out the devil."

Since these ideas about sickness and death and the fate of the dead are part and parcel of the mental stock of primitive man, and have remained unchanged in essence, and but slightly altered in form, to this day, their importance cannot be overrated. Without them, futile are all attempts to explain the meaning of customs that have ruled nearly every act of life of many peoples, and that still rule a large part of the life of all peoples. With them, we read, as in open book, what mean the scattered relics of feastings at the grave; the modelling of the tomb upon the lines of the house; the decoration and furnishing of it as a place where the ghost might eat, drink, and be merry, and enjoy the flowers that garlanded it and the libations poured upon it. And only with knowledge of these ideas can we follow the steps by which the transition was made from grave to altar, and from altar to temple. For the burial-places, where feasts of the dead were held, became altars whereon sacrifices were offered to the ancestral spirits, worship of whom is a leading element under various forms in the religions of every race and age, whether among barbaric or civilised peoples. The altars, each containing its relics of the dead, were the nuclei of temples from the rude stone circle upward, through many gradations, to the church of St. Peter in Rome, wherein the saint's bones are said to lie.

The stone structures which, in more or less ruined state, are scattered in thousands over both hemispheres, are the special feature of Neolithic times. They are reducible to three classes: 1. Single upright stones called *menhirs* (from Celtic *maen*, a stone; and *hir*, high), which may have been commemorative of events, or set up as monuments of the dead. 2. *Dolmens* (from Celtic *daul*, a table; and *maen*), consisting of three or four rude stones on which another stone is placed, forming a chamber; a structure of wide-spread distribution, and of which "Kits Coty House," near Aylesford, in Kent, is a notable example. 3. *Cromlechs* (from Celtic *crom*, a circle; and *llech*, a stone), circles enclosing barrows and dolmens, or standing by themselves. Like the long and round tumuli, they are modelled on the primitive dwelling, the ring of stones corresponding to the fence round the Neolithic "hut-circles," or to the stones which prop up the turf-covered houses of the Lapps. Misnamed as Druids' Circles, Giants' Graves, Odin's stones; the home of legends of surprising deeds wrought by the Devil; of myths of ogres and dwarfs; of Sabbath-breakers and infidels turned into stone, these venerable cromlechs are invested with a mystery which deepens their charm. If not themselves the objects of worship they stand in near alliance with, to us, that strangest of primitive idolatries, the worship of stocks and stones, than which none has had a wider range. The cause of this expression of the religious sentiment lies in the universal barbaric tendency to confuse persons and things, and to attribute, through superficial and seeming likenesses, virtue or power to lifeless objects.

Borrowing what is the best illustration of this, namely, the practice of barbaric folk as observed by travellers, we are told that in Melanesia, when a man sees a fantastic-looking stone, one, say, shaped like a breadfruit, he buries it near a breadfruit tree. Perhaps, through a favourable season, he gets an extra crop, whereupon he is satisfied that the stone has a spirit in it which has made the tree yield better. Hence the stone becomes sacred, an object of worship, an heirloom in the family. If, in the same islands, a native comes upon a big stone with a number of little stones underneath it, like a sow with her litter, he will make an offering to it, in the belief that it will bring him pigs. Mr. Codrington, who is the authority for these examples, says that sacred stones abound, especially in all sacred places, in the Solomon Islands; food is laid on them to secure good crops, cooked fish being used if a large catch is wanted, in accordance with the barbaric idea that like produces like, however far-fetched the similarity may be.

In this, ancient and modern meet together. The anointing of the sacred stone at Bethel by Jacob is paralleled by the Society Islander, who smears a basaltic rock with oil; and the reproach of the later Isaiah against the drink-offerings of the Jews to the "smooth stones of the valley" shows them at the level of the natives of the New Hebrides, who worship water-worn pebbles. And if, as has been seen, celts and arrow-heads have been looked upon as of celestial origin, or as the work of superhuman beings, aërolites or meteoric stones have, with better reason, been regarded as falling from heaven. Such is the "Black Stone" of the sacred Kaabah at Mecca, which is reported to be an aërolite; such too, perhaps, is the pyramid-like stone worshipped by the Hindus at Jagannath, as also the stone which, falling near a sacred place, was worshipped in Mexico as the son of divine parents.

Jews, Greeks, and Romans were on a par with the Hindu villagers of to-day in the worship of standing stones; unwrought stones had place of honour in classic temples, representing the greater gods and "receiving care and decoration as well as worship." In the Temple of Heaven at Pekin seven unhewn boulders guard the fortunes of the Imperial dynasty; and with the ceremonial seats hewn in the solid rock for Mexican rulers may be classed the stone of Scone which is underneath the seat of the coronation chair in Westminster Abbey. It is called the *Lia Fail*, or "Stone of Destiny," and is said to have been the stone which Jacob used for a pillow when he slept at Bethel, and had sweet dreams of angels. Somehow it reached Ireland (home of many a tradition), and was finally brought to Scone. Another old relic, the London stone in Cannon Street, connects the sacred boundary-stone with Hermes or Mercury, among whose many characters was that of a boundary god in the shape of an upright stone. In their boundary-stone festival—the Terminalia—the ancient Romans honoured one of the oldest and most sacred of their divinities. For the well-being of a people is dependent on the cultivation of the soil, which involves measurement and boundary-marks: hence the importance of these, and their divine protection. The persistence of stock and stone worship in Europe is shown in the succession of decrees issued against it down to the reign of Cnut, in the eleventh

century. About the seventh century the Council

FIG. 70.—STONEHENGE.

of Nantes "exhorts Bishops and their servants to dig up and remove and hide in places where they cannot be found those stones which in remote and woody places are still worshipped, and where vows are still made," and four hundred years later a statute of Cnut forbids the "barbarous worship of Stones, Trees, Fountains, and of the heavenly bodies."

VI. *Stone Circles*. To return to the cromlechs. Of these the largest relic in this island is near Abury (or Avebury) in Wiltshire. Aubrey, an antiquary of the seventeenth century, describes it as "exceeding in greatness the so renowned Stonehenge as cathedral doeth a parish church." But it is so ruined as to be scarcely distinguishable to the unpractised eye seeking for fragments of its circles within circles amongst the barrow-strewn ridges. Indeed, barely a score of stones out of hundreds now remain. Most of them have been used to build the modern village which stands within the ramparts that surrounded the outer ring of stones.

Later in time is the famous Stonehenge, one of the most impressive ruins in the world, both as regards character and situation. Standing before this mute, desolate pile, while we can only speculate on what strange rites, what bloody sacrifices, have been performed within it, we may feel sure that it represents a stage of religious culture similar to that which the foregoing examples illustrate. Although it is a confused ruin, the original plan has been clearly traced. Briefly stated, it consisted of two concentric circles enclosing two ellipses both open at the north-east end, the whole being enclosed within a low earthen rampart and ditch.

The outer circle was composed of large squared and roughly hewn pillars with imposts morticed

FIG. 71.—TRILITHON, STONEHENGE.

into them, forming an unbroken colonnade. They are of a sandstone which occurs in the neighbourhood, and are locally known as "sarsens," from a myth that they were "Saracen" or Paynim warriors who had been turned into stone. A similar legend has gathered round the Rollright Stones, a cromlech in Oxfordshire, the standing stones of which are said to be a petrified king and soldiers. The inner circle consists of undressed stones called "blue stones," which must have been brought thither, since they are not found nearer than North Wales or the Channel Islands. The ellipses are formed of five trilithons, or two huge upright stones supporting an impost or lintel. These were arranged in horse-shoe form, and are of the same material as the stones of the outer circle. A smaller ellipse of "blue stones," but without imposts, stands within them, and within this double "horse-shoe" is a large flat slab called the "altar stone." The circles were approached by an avenue in which stands a large unwrought stone known as the "Friar's Heel." The interesting feature in connection with these two last-named stones is that on the longest day of the year the sun rises immediately over the "Friar's Heel," his rays falling on the "altar stone," and that on the shortest day the sun rises and sets directly over some small stones beyond the outermost circle. This has led to the theory that Stonehenge was a seat of sun worship, but the various suppositions about its origin and exact purpose cause us to fall back on the cautious remarks which Pepys sets down in his Diary after visiting the place in 1668. "God knows what their use was. They are hard to tell, and may yet be told." Still, we have made no small advance in knowledge of the significance of ancient monuments since Pepys' day, and, with the three hundred tumuli within a radius of three miles from Stonehenge, may feel little doubt that it was sepulchral in origin. Recurring to the unquestioned relation of the dwelling of the living to the tomb of the dead, we may see in the surrounding earthwork the village rampart; in the avenue—dwarfed type of the extensive ruins at Carnac, in Brittany—the underground gallery leading to the pit-dwelling; and in the circle the enlargement of the ring of stones which surrounded or supported the bee-hive-like hut. The exploration of the area within the pillar-stones has not been complete, and the fragments found—bones of animals and sherds of rude pottery—have added little to our knowledge. If any support could be found for the theory that a tree—the sacred oak, a supreme object in the nature worship of Neolithic people in Europe—stood within the central sanctum, some links between the old temple and the Druids, about whom much nonsense has been written, might be established.

Before passing from the great monument of our Wiltshire downs, with their dykes and barrows, and relics of Old Sarum and the like, the significance of the sun's rays in striking the "altar stone" at the summer solstice invites brief comment. Possibly we have in this a relic of sun-worship, but there is no warrant for Sir Norman Lockyer's fantastic theory that Stonehenge was built by priest-astronomers from the

East, because, as he argues, Egyptian temples were likewise so designed that their direction was governed by alignment with prominent stars, so that the light flashed down them and illuminated the figure of the god which stood in the sanctuary at the end of the long axis. The direction of Egyptian temples has sufficing explanation in the lie of the Nile Valley and in the conformation and nature of the soil. Sun worship possibly survives in what is called the "eastward position," that is, the "orientation," or building of churches so that the chancel points to that part of the east in which the sun rises on the day of the saint to whom the church is dedicated. It also explains the burial of the dead sitting with face to the east, although often in the opposite direction, looking westward, where many a barbaric folk has placed the home of souls. As Dr. Taylor remarks, "it is not to late and isolated fancy, but to the carrying on of ancient and widespread solar ideas, that we trace the well-known legend that the body of Christ was laid with the head towards the west, thus looking eastward, and the Christian usage of digging graves east and west which prevailed through mediæval times is not yet forgotten. The rule of laying the head to the west, and its meaning that the dead shall rise looking toward the east, are perfectly stated in an ecclesiastical treatise of the sixteenth century."

Stone circles extend, with certain gaps, from Scotland to the Antipodes. The largest example north of the Tweed is the cromlech at Stennis, in Orkney, near which stood a holed stone, celebrated as the betrothal place of lovers who, in clasping hands through it, regarded the vow as especially binding. Some writers conjecture that the wedding ring, as a piece of perforated metal, has its origin in customs like the foregoing. Be this so or not, every boy who carries a lucky-stone in his pocket represents the superstitions which have attached themselves to stones with a hole in them, whether the hole be large enough to pass a child through for cure of some disorder, or the stone small enough to be used, as among rustics in Suffolk and elsewhere, as a charm against nightmare; or tied to the key of a stable-door to prevent the witches from stealing the horses. The standing stones at Carnac, in Brittany, were arranged in the form of eleven roughly parallel avenues, which must have originally extended several miles. Whether they led to any central monument is unknown. Their purpose remains a mystery, like that of the storied cone-shaped towers in Sardinia, called *nuraghe* (although Mommsen calls these "sepulchral"), and the cyclopean "talayots" of Minorca.

Cromlechs are not uncommon in Northern Arabia and the peninsula of Sinai, while, as showing that the old motives still prompt their erection, the Khasias of Bengal, "among whom only funeral ceremonies are of any importance," raise undressed stones of gigantic size as monuments, singly or in rows, or circles, supporting one another like those of Stonehenge, which they rival in dimensions and appearance. Aristotle tells us in his "Politics" that the Iberians were in the habit of placing as many stones round the tomb of a dead warrior as he had slain enemies, and some Australian tribes are reported to place pillar-stones round the burial-place of famous brave or chief. If no other evidence existed, that which is supplied by the rude stone structures in all parts of the globe suffices to link the barbaric peoples of every age in one common chain of ideas, and of practices which are the outcome of those ideas. And the same chain lengthens till it touches the higher races, in whose finished products the raw material can be traced. The menhir is the rude original of the columns of Trajan and of Nelson; the dolmen of the sarcophagus of Ramses and the tomb of Wellington; and the cromlech of our Temple church, which was built by the Templars in the twelfth century in recollection of the round church of the Holy Sepulchre, at Jerusalem. Between Stonehenge and the fair cathedral whose spire we see as we return to Salisbury the chain of continuity is complete.

Although perhaps not the most consistent, this is the more convenient place to refer to the remarkable earth mounds which exist in thousands over the central parts of North America, especially in the basin of the Mississippi. For although the contents show that their builders were acquainted with metals in the shape of copper, the primitive way in which they worked a material found in hard lumps in a native state, namely, by simply hammering it into the required shape, evidences their being in the transition stage between the Ages of Stone and Bronze. The mounds are divisible into three or four groups: military or defensive; sepulchral or temple; and animal. As with the remains which witness to the relative succession of "Iberians" and "Celts" in Europe, so these American remains yield traces of a long-headed race and of a later and conquering round-headed race; while, to complete the interesting similarities, the finds of rudely-chipped stone implements point to a Palæolithic folk, who perhaps battled with the megatherium (a huge mammal allied to the sloth) when his fellow flint-chipper was killing the mammoth in Europe. The vast extent of some of the fortifications is shown in the remains of Fort Ancient, in Ohio. These are "nearly a mile in length from north to south, with more than 20,000 feet of wall, more than five miles of terraces, and more than ten miles altogether of artificial work." They may have formed part of a connected line of defence as far as New York. The burial mounds outnumber the military by tens of thousands, most of them containing one body buried in the contracted posture referred to already, and the others only the burnt ashes of the corpse. Some of these mounds were modelled on the plan of the dwellings; others are probably dwellings which have been converted into burial-places. Both sepulchral and temple mounds yield traces of barbaric sacrificial rites. Among the distinctive relics are immense numbers of carved and ornamented pipes, which appear to have been used in connection with offerings to the gods, for there is little doubt that the "fragrant weed" was burnt at their altars, and originally applied only to sacred purposes. Strangest of all are the "animal mounds," shaped in low relief on the surface of the ground as figures of men, birds, and reptiles, extending in some instances to a length of several hundred feet. One of the largest, called the "Great Serpent," starts from an oval enclosure at the end of a promontory by a tributary of the Ohio. The jaws are distended, "as if in the act of swallowing what, in comparison with the huge dimensions of the head, is spoken of as an egg, though it measures 160 feet in length. Conforming to the summit of the hill, the body of the serpent winds back for 700 feet in graceful undulations, terminating with a triple coil at the tail. The figure is clearly and boldly defined, the earth-wrought relievo being upwards of five feet in height by 30 feet in base at the centre of the body, and diminishing towards the head and tail. The entire length, following its convolutions, cannot measure less than 1,000 feet. On either side of the serpent's head two small triangular elevations extend, looking on the ground plan like external gills, but they are so much obliterated as to render their original form uncertain." Speculation was for long rife as to the race who had left these curious structures behind, but there is no longer any question that "the mysterious 'Mound-builders' were the ancestors of the Red indians. Every relic of their supposed civilisation corresponds with something that can be clearly traced to ordinary

Indian hands; and traditions have here and there survived attributing the great earthworks of the Ohio to the red man's ancestors. These traditions ascribe their construction to an age not very remote, when those ancestors were more numerous and prosperous than in historical times, and when they were liable to attack in the course of southward migrations which were taking place on the Pacific side of the Continent. The history of the 'Mound-builders' thus merges in the general story of the American aborigines."

The barbaric mind, with its belief in spirits dwelling in lifeless things, in big stones as the parents of little stones, and so forth, seems to reach a higher plane when it conceives of a life shared in common by man and animal and plant, and thereupon frames its myths—real enough to itself—of human descent from trees and animals, rather than from stones. The passage from this to the worship of plants and animals is then a short one, if, indeed, the steps be not taken together. Be this as it may, the belief in such descent is found among barbaric folk everywhere, and must have had its germs in the mind of primitive man. It has led to some curious customs, such as not eating the animal or plant which is the "totem" or "clan-mark" of the tribe, and in not marrying a woman of the tribe because she bears the same name as the man. The subject, full of interest as it is, would lead us too far afield, and is referred to here as possibly throwing light on the carving of the soil into fantastic "animal' mounds. For totemism survives in its most primitive form among two races: the black fellows of Australia and the North American Indians.

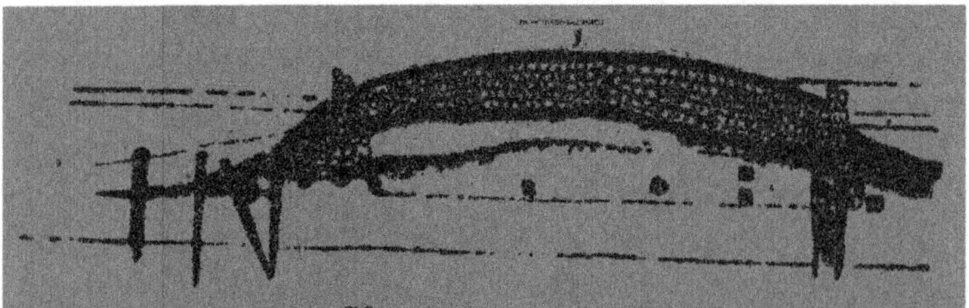

FIG. 72.—IRISH CRANNOGE.

The word itself is derived from the Algonquin "dodaim" or "dodhaim" = "clan-mark." Man everywhere makes images of his gods as he

FIG. 73.—IDEAL RESTORATION OF A SWISS LAKE DWELLING.

conceives them, and the "Great Serpent," with its allied figures, may be the Red Indian's tribute to the creatures whom his fancy and his fears have promoted from ancestors to the rank of deities.

VII. *Remains found in Lake-Dwellings.* A welcome addition to our knowledge of Neolithic times was made through the discovery of remains

FIG. 74.—LAKE DWELLING. (Cameron's "*Across Africa.*")

associated with defensive structures—as 'crannoges'—built on islands, both natural and artificial; and, more especially, with dwellings on piles driven into the bed of lakes. This mode of life, obviously chosen as a protection againt foes and wild beasts, is still extant, "pile-dwellings" being common in Central Africa, South America, New Guinea, Borneo, and elsewhere. Venice, the dreaming city "that is always putting out to sea," which was founded about fifteen

FIG. 75.—TOWN OF BRUNEI, BORNEO, BUILT ON PILES.
(*Clutterbuck.*)

centuries ago by refugees from the mainland, is an example on a large scale. History records the use of such dwellings in Greece and Asia Minor; a representation of one is sculptured on Trajan's column; and the traces of a pile-dwelling were found in Moorfields, near London Wall. An old writer of the fourteenth century, quoted by Dr. Keller in his classic work on "Lake Dwellings," speaks of the Apamœan lake in Asia Minor "as commonly called the lake of the Christians, because it is inhabited by Christian fishermen, who live in wooden huts built upon piles." But the earliest notice of such dwellings is given in Herodotus, who thus describes the aquatic life of the ancient Pœonians on Lake Prasias, the modern Lake Takiros, in Thrace.

"They who inhabited Lake Prasias were not conquered at all by Megabazos. He sought indeed to subdue the dwellers on the lake, but could not effect his purpose. Their manner of living is the following: a platform fastened together and resting upon tall piles stood in the middle of the water of the lake, with a narrow approach to it from the mainland by a single bridge. At first the piles were no doubt fixed in their places by all the members of the community working together, but since that time they continue to set them by observance of this rule; that is to say, every man who marries brings from the mountain called Orbelos three piles for each wife, and sets them as supports; and each man takes to himself many wives. And they have their dwelling thus, that is, each man has his own hut upon the platform in which he lives, and a trapdoor leading through the platform down to the lake; and their infant children they tie with a rope by the foot, for fear they should roll into the water. They feed their horses and their beasts of burden upon fish;

and of fish there is so great quantity that if a man open the trapdoor and let down an empty basket by a cord into the lake, after waiting quite a short time he draws it up again full of fish." Herodotus lived two thousand four hundred years ago, but fishermen still have their wooden huts on Lake Prasias. The history of nearly all the relics of pre-historic man repeats itself in the tardy or ungracious recognition of their meaning and, as it has always turned out, their high value. Although as far back as 1829 artificially shaped piles and other remains were discovered when the harbour at Ober Meilen, on the Lake of Zurich, was being deepened, they excited little inquiry. A quarter of a century later the dry winter caused the lakes and rivers of Switzerland to fall much below their usual level, and the inhabitants of Ober Meilen took advantage of this to reclaim the exposed land, raising it with mud dredged from the neighbouring shallow water. In removing the mud the workmen came upon a number of deeply-driven piles of oak, beech, fir, and birch, which some fishing folk, whose nets had been often entangled in them, had mistaken years before for submerged forests. Around the piles were found large numbers of primitive weapons, both of stone and bone, fragments of coarse pottery (of the potter's wheel, the invention of which is claimed by the Greeks—there is no trace), charred wood, and other burnt materials, the examination of all which satisfied Dr. Keller that the piles had supported a platform on which huts had been built, and that the whole structure had been ultimately destroyed by fire. The researches to which this discovery gave impetus have been carried on since 1854 to the present time, and established the fact that a large population lived in this aquatic fashion in Central Europe and the British Isles, not only through Neolithic times, but also through the Ages of Copper, Bronze, and Iron, and in Switzerland

FIG. 76.—POTTERY FROM AUVERNIER, ETC.

until the first century after Christ. Above two hundred of these habitations are known to have existed in Switzerland, the Lake of Neuchatel having furnished one-quarter of the whole. And as each village contained on an average about three hundred huts, the population was not inconsiderable. Choosing a sheltered and protected spot, the settlers cut down trees with their stone axes, and sharpened them with fire: a long and arduous labour, some forty to fifty thousand piles being needed for a small village.

These were then driven into the bed of the lake, about two or three hundred feet from the shore, till they were all at the same level, when wooden planks were fastened to them with pegs to make the platform. On this were raised the square or oblong-shaped huts, each large enough to contain a family. They were made of woven twigs plastered with clay, the roof being thatched. Between each dwelling were placed the cattle-pens, sheepfolds, and pigsties, a gangway being built between the village and the shore, for the use of the cattle when driven to pasture, and for general access to the mainland. The

FIG. 77.—"DUG-OUT" CANOE, MEININGEN.

lake-dwellers fished, either from the stage or in their "dug-out" canoes, hunted, tilled the soil, and tended the cattle.

Some of the sites contain remains belonging to different periods, and this mixture of stone and metal relics complicates any attempt at clear arrangement. But without making the blunder of forcing every class of objects into one of the three "Ages," as to the over-lapping of which the Lake-dwellings are among the numerous examples, there are sufficient distinctions among the several groups to assign some to the Stone Age (in the Lakes of Bienne and Neuchatel), others to the Bronze Age, and a

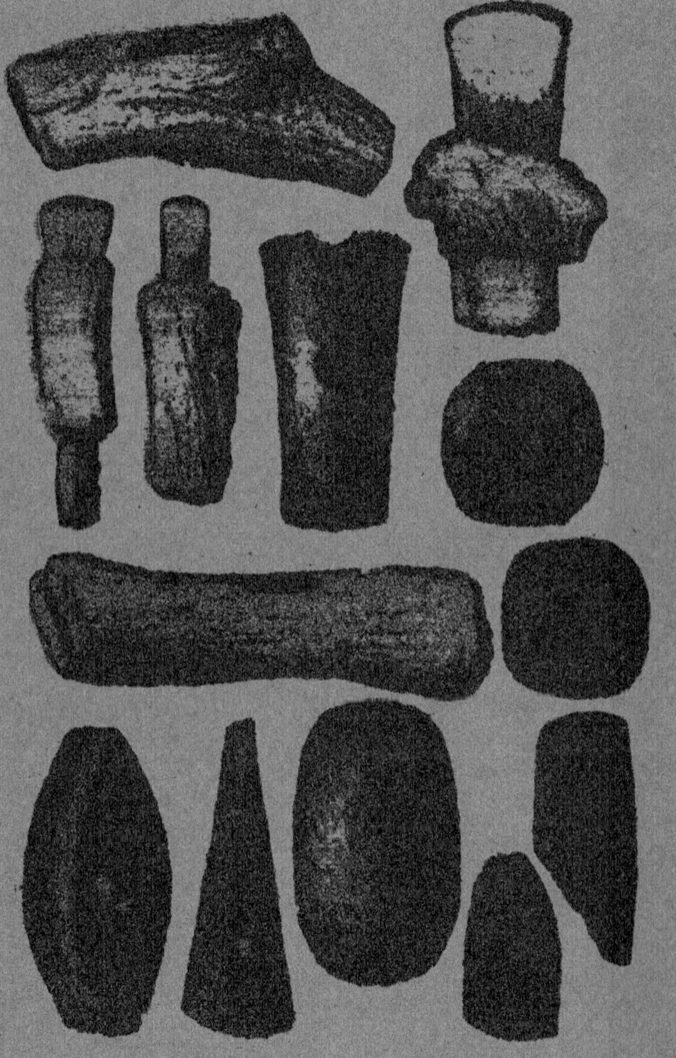

FIG. 78.—STAG'S-HORN HAMMER, CELTS, STONE HAMMERS, CORN CRUSHERS, ETC. FROM LAKE DWELLING, MEILEN.

very few to the Iron Age. The pile-dwellings of East Switzerland are the earlier, and those of West and Central Switzerland the later, the number becoming fewer as civilisation advanced, rendering life and property more secure, and encouraging the settlers to remove from the water to the land. The modern Swiss châlet is probably a copy of the lake-dwelling. Buried in the soil with fragments of the huts, a rich harvest of relics has been gathered, wherewith, aided by our knowledge of modern aquatic modes of life, a fairly complete sketch of these Neolithic folk has been made possible. Arrow-heads, lance-heads, and polished stone axes, fragments of rude nets and of "dug-out" trunks, tell of a hunting and fishing population, while other relics evidence advance to the shepherd and farmer stage. The remains at Robenhausen, which cover three acres, show that "the litter for the cows was chiefly of straw and rushes, and for the sheep, pigs, and goats, of sprigs of fir and twigs of brushwood." The bones of these animals, and of fish caught in the lake, are mingled with those of animals of the chase. And further evidence of progress is at hand in the stores of grain, of apples, pears, seeds, and berries; in the loaves of meal with the mortars or mealing-stones for crushing it; in plaited flax and woven nets; in earthenware weights for the loom; in fragments of leather, and even in wooden lasts for shoes.

Weaving and plaiting are among the earliest arts. The need of securing things together, or of otherwise strengthening them, led to binding, fastening, and sewing. The wattlework hut, with its roof of interlaced boughs, the skin sewn by bone needle with sinews, the matted twigs or fibres, are all the rough beginnings of an art which we can trace step by step. In the twisting of fibres or hairs by rolling them between the palms of the hand we have the original of the spinning-wheel and the steam-driven cotton spindles; in the roughest plaiting we have the first hint of the most superfine cloth. And these flaxen hanks or skeins spun or plaited into cords, these nets and woven fabrics of the lake-dwellings, tell of a settled life in which the "man delved and the woman span."

There, too, were the oldest farmers and stock-breeders known to us. The number of animals partly or wholly domesticated by them shows that the process of taming must have begun ages before their arrival in Central Europe; perhaps in the regions lying between them and the rude ancestry who had tamed only the dog, a creature whose social instinct made that task easy. The process of selection, which ensures domestication, must have ruled from pre-historic times. Mr. Galton's description of it as in operation among barbaric tribes nowadays enables us to picture the nomads of the past slowly bringing wild ox, wild boar, and other untamed species into subjection. He says, "the irreclaimably wild members of every flock would escape and be utterly lost; the wilder of those that remained would be selected for slaughter whenever it was necessary that one should be killed. The tamest cattle—those which seldom ran away, that kept the flocks together, and those which led them homeward—would be preserved alive longer than any of the others. It is, therefore, these that chiefly become the parents of stock and bequeath their domestic aptitudes to the future herd. I have constantly witnessed this process of selection among the pastoral savages of South Africa. I believe it to be a very important one on account of its rigour and regularity. It must have existed from the earliest times, and have been in continuous operation down to the present day."

Although we speak of man as first a hunter, then a shepherd, and then a farmer, this only roughly indicates his advance from the wandering to the settled state. And these stages are not universal, some races being agriculturists, in a rude sort of way, from the outset. Agriculture may be defined as the domestication of plants, or the selection of wild varieties which man discovers to be good for food, although, as contrasted with wild animals, many of them can be more easily utilised without domestication. The beginnings of agriculture are too remote for the primitive processes to be followed; but they can be observed among barbaric races nowadays and, in large degree, in remote parts of civilised countries, where the old methods are little altered, and the steam plough has not superseded the crooked branch. Interest gathers round the question of the sources whence many of our cultivated varieties have been derived, and, as usual, the East has been pointed to as their native home. But although this is true of some, examination of deposits in Europe has shown that others were indigenous. The connection between agriculture and religion has been unbroken from pre-historic times. Grant Allen, in an ingenious paper published in the *Fortnightly Review* of May, 1894, argues that cultivation began with the accidental sowing of grains upon the tumuli of the dead, because it would not occur to primitive man to save the seeds and bury them in the soil on the bare chance of their sprouting and producing seeds after their kind. Seeing that this happened on graves, he would conclude that the grateful ghost beneath had thus repaid him an hundredfold the offerings he had made. Step by step the land about the tumuli would be cleared and cultivated, but always with the provision of a corpse, or fragments of a corpse, buried in the soil as the condition of fertility. "With the gradual mitigation of savagery an animal sacrifice was often substituted for a human one; but the fragments of the animal were still distributed through the fields with a mimic or symbolical burial . . . the idea of the crop being a gift from the deified ancestor or the divine human victim being kept up in the common habit of offering the first fruits to the dead, or to the gods, or to the living chief, their representative and descendant." The theory is illustrated by striking examples of customs connected with the worship of the dead, and, in so far, has much to commend it. But seeds sprout on other places besides graves, and it needed no very careful observation on the part of primitive man to note so common an occurence, while he would attribute it to the spirit which, as has been shown, he believed to dwell in everything. All around him was the magic outcome of life, whether from the womb, or the

FIG. 79.—OJIBWAY GOD OF THE GRASS. (*Dorman.*)

egg, or the seed: the ceaseless succession of birth, growth, and death; and after that, and evermore, birth, growth, and death. Throughout the religions of barbaric races we find worship of the spirit of each plant, as the "grass-spirit," "corn-spirit," the "maize-spirit," the "tobacco-spirit," and, with these, worship of sun-god and rain-god, without whom the plants would not grow, while, chiefest of all, is the great Earth-mother, round whose worship the popular rites of every people have gathered. As Tocitzin among the Mexicans, Demeter among the Greeks, and Maia (the early earth-worship surviving in May Day and kindred festivals) among the Latins; prayers were offered to the fruit and corn-giving "mother of all." And, as showing how in naught else so much as in religion the belief of primitive man in spirits in everything survived, it was held in old Rome that no bounteous harvest would be granted unless the farmer invoked with Mother Earth and Ceres "the spirit of breaking up the land, the spirit of ploughing it crosswise, the spirit of furrowing, of ploughing-in the seed, of harrowing, of weeding, and of reaping, the spirit of carrying the corn to the barn and the spirit of bringing it out again.

As bearing on the stages of culture represented in settlements which date from the Neolithic Age and extend to the Christian Era, it is interesting to note that in the oldest series, the remains of the stag predominate over the ox, of the goat over the sheep, of the wild boar over the domestic hog, and of the fox over the dog. In settlements referred to the Bronze Age the dog predominates over the fox, the domestic hog over the wild boar, and the sheep over the goat. At the Steinberg, identified from a few iron implements met with there as one of the later villages, numerous bones of the horse are found—an animal whose remains are extremely rare in the earlier series of dwellings. The absence of remains of the hare from the *débris* is curious, as perhaps showing that the prejudice against eating its flesh, which Cæsar speaks of as existing in Britain, extended to Central Europe—a prejudice possibly due to "totemism." The authority of Mr. Elton may be cited on this matter. In his "Origins of English History" it is remarked that "It seems reasonable to connect the rule of abstaining from certain kinds of food with the superstitious belief that the tribes were descended from the animals from which their names and crests, or badges, were derived. There are several Irish legends which appear to be based on the notion that a man might not eat of the animal from which he or his tribe was named. . . . The hare is now an object of disgust in some parts of Russia, as well as in Western Brittany, where not many years ago the peasants could hardly endure to hear its name. The oldest Welsh laws contain several allusions to the magical character of the hare, which was thought to change its sex every month or year, and to be the companion of the witches who were believed to assume its shape."

VIII. *Origin of the Lake-Dwellers.* We have travelled a long way from the Thames and Somme valleys and the Dordogne and Devonshire caverns, reaching the Swiss lake shores not only by crossing seas where, at starting, none existed, but also gaps which only legend and tradition fill. Gaps also, though narrower, between the men of the kitchen-middens and the men of the pile-dwellings, all leading to the question of the connection of peoples in prehistoric times.

The remains of the lake-dwellings indicate different stages of civilisation in the *same* sites. Around some of the piles, stone, bronze, and also iron articles occur together, showing that settlements founded in the early Neolithic Age passed without apparent break to the metal ages. Like evidence has been obtained from the pile-dwellings in the valley of the Po, although these did not survive beyond the age of bronze. Thus the dwelling opposite Peschiera, on the Lago di Garda, was founded in the Neolithic Age, and was in continuous occupation through the age of copper to the age of bronze. The remains of the settlement in the Lake of Frinon exhibit two successive beds, the oldest being wholly neolithic. The villagers were mainly in the hunting stage, the bones of stags and wild boars being abundant, while those of oxen and sheep are rare.

There are no remains of cereals of any kind, but great stores of hazel-nuts were found, together with acorns, some of them adhering to the inside of the pipkins in which they had been roasted for food. The settlement seems to have been burnt, and then after a time rebuilt, the newer relic-bed containing numerous flint chips and a solitary bronze axe. Cereals are still absent, although acorns, hazel-nuts, and cornel cherries are found. But the pastoral stage had plainly been reached, since the bones of the stag and the wild boar become rare, while those of the ox and sheep are common.

The alternative, therefore, is that the lake-dwellings were either inhabited from their foundation to their disappearance by the same race which gradually passed from the use of stone to that of metal, acquiring this by barter; or that they were invaded by a superior race armed with bronze weapons: a race which conquered the lake-dwellers, drove them into the wilds, or perhaps exterminated them (that is, if any race in Europe ever was exterminated), and occupied their settlements. Until recently it was taken for granted by well-nigh every writer on the subject that the latter alternative was the true one. Everything came in the train of the Wise Men from the East: the birthplace of mankind, the cradle of civilisation. Ancient traditions lay at the base of this unchallenged theory, and confirmation of it appeared certain in the interesting discovery of a close family relation between the languages spoken by the leading peoples of Europe and by the dominant peoples of India and Persia. The dissection of words, and of the roots from which they spring, evidenced that from the mother "Aryan" or "Indo-European" tongue there had descended the rich and vigorous languages in which are written the Vedas, or sacred books of the Hindus; the Avesta-Zend, or sacred books of the Pârsis; the Iliad and Odyssey; the Eddas; the Divina Commedia; and the plays of Shakspere. Also that the people who spoke this mother-tongue lived somewhere east of the Caspian Sea in a fairly advanced state of civilisation at a period when Europe was a jungle inhabited by the men of the kitchen-middens and the earliest lake-dwellings. Sanskrit, the language in which the Vedas are written, was adjudged to be nearest of all to the original tongue, and the oldest member of the family of speech derived from it. It was therefore argued that the primitive home of the Aryans (using this term not as a race-name, but to denote people speaking allied languages) was in Central Asia, whence swarmed, one after another, the various tribes: some southward, through the passes of the Hindu Kush into India, some into Persia; others by routes north of the Caspian Sea into Russia and Central Europe, or, south of that sea, into Greece and Italy. The first to arrive were the Celts, who, in the course of time, were driven westwards to the extreme corners of Europe by the forefathers of the Germans and Slavs; the tribes that took a more southerly path being the ancestors of the Greeks and Romans. Thus was it all made clear that peoples now as diverse as the Hindu and Icelander, the Greek and German, the Persian and Englishman, and the Russian and the Irishman, are the offspring of a mother-race that nurtured them on the high-lands of Asia, and sent them forth, one by one, as more elbow room was needed, to conquer and to civilise. And, therefore, it was to these Aryans that the introduction of a higher culture among the lake-dwellers, or such of them as according to old notions, were not exterminated, was due.

All this sounds very plausible—indeed, has a dash of the poetic in it, but it is, as idylls often are, largely a work of imagination, so far as the cradle or primitive home of the Aryans is concerned. It rests on the fallacy that identity of language implies identity of race. Whereas, as Professor Rhys says, "skulls are harder than consonants," and while the hall-mark of race is never wholly rubbed out, its language is as unstable as the sea-waves. Numerous causes operate to bring about change; conquering races, or intermixture with races of superior culture, effect it, often sweeping away the old language, and imposing a new language upon an intellectually inferior, even though physically superior, race. Languages prove social contact, but not blood-alliance. The Jews, while keeping themselves distinct as a race, have adopted the language of the various countries in which they have settled; the Cornishmen and Bretons are largely of the same blood, but their language is different; and although the French and Italian languages point to descent from a common tongue, it does not follow that Frenchmen and Italians are the offspring of common ancestors.

But, even if it could be granted that language is a test of race, the evidence that Sanskrit and Zend, as the speech of the ancient Asiatic Aryan, are most nearly allied to the primitive Aryan tongue, is disputed. Sanskrit, the language of the Hindu scriptures, and Zend, the language of the Persian scriptures, are no longer spoken, whereas the Icelandic and Lithuanian tongues are not only still spoken, but have the best claims to represent the Aryan mother-tongue; those, for example, who speak Lithuanian "being probably the direct descendants of those who spoke it two, or possibly three, thousand years ago."

Language alone, then, has been made the basis of a theory of successive immigrations of Asiatics from warmer regions over frozen steppes or across tempestuous inland seas, followed by the extinction of non-Aryan Europeans. Now we find the main body of Aryan-speaking peoples in Europe, and only a small-detached body in Asia—the main trunk here, two severed branches there. The conclusion therefrom is not that the larger mass broke away from the smaller, but the smaller from the larger. "The European Aryans form a closely united circular chain of six links, but there is one vacant place—one link is missing from the chain. This missing link is discovered far away in Asia, where we find the Indo-Iranians, who are very closely united with each other, but whose affinities with the European Aryans are chiefly with the Slavs on the one hand, and with the Greeks on the other. They clearly constitute the missing link in the chain."

It is highly probable that the races who spoke the Aryan mother-tongue were scattered over Europe at the not very remote period when Asia Minor was joined to Europe at a point where the Bosphorus now flows. The waters of the Black Sea were then united to the Caspian and Aral Seas and to Lake Balkash, forming one vast inland sea, which overflowed northward through the basin of the Oti into the Arctic Ocean. With the wearing away of the isthmus between Europe and Asia Minor the waters of the Black Sea poured through what is now the Dardanelles to the Mediterranean, and, with the warmer climate and drier air, brought about by the slow upheaval of Siberia, the Caspian, the Aral, and the Balkash waters shrunk under more rapid evaporation to their present size, laying bare the wide, salt-laden wastes that now spread between them. Across the broad plains there passed eastwards some of the Aryan tribes, ancestors of Hindus and Persians.

Of course language is a test of race among people who have had no foreign tongue imposed upon them; and, to the extent that the same things are called by the same names among people now widely scattered it is a proof of former contact, and, in high degree, of blood-relationship. And as giving the clue to what were their thoughts and material possessions, it helps us to estimate what stage of culture they had reached *before* they separated. Where the same thing is called by different names in the separated branches, we conclude that it became known to each independently *after* the separation. Where the name of anything is present in one branch and absent in the other, we draw the same conclusion. A few examples will show the bearing of this on our knowledge of the primitive Aryan civilisation, and on the relative periods when contact between the several peoples was more infrequent until it finally ended.

The beech does not grow east of a line drawn from the south of Scandinavia through Poland to the Crimea and Caucasus. The name for it is common to the Celtic, Teutonic, Greek, and Latin languages, but has been borrowed from the Teutonic by the Slavonic, and is absent from the Indian and Persian languages. The tree is unknown in Central Asia, and so it would follow that the Eastern branch of the Aryans had migrated thither after the period when the beech had begun to replace the oak in Denmark. And this was not till the "Age of Iron."

Pastoral people divide the year into two seasons—the one when the cattle are fed in the open, the other when they are housed in the stall. And among the Aryan languages we find common words for summer and winter, winter being the season most sharply defined; "both these," as Tacitus says in his "Germania," "have a meaning; but the name and blessings of autumn are alike unknown." This implies the absence of an in-gathering season, and therefore of a definite agricultural stage, as indicated in the terms used later; *e.g.*, in the German *herbst*, or "harvest-time." The related words for the sea appear only in the European group, but the names for rowing and boat are identical in both that and the Asiatic group. The name for the oyster is the same in the European branch, but it is absent from the Asiatic. The names for the dog, cow, sheep, and horse (which animal was eaten, but, perhaps, not yet then domesticated) are alike, and the possession of cattle among the undivided Aryans is shown in the common name, which goes back to a root meaning "capture." The domesticated animals were denoted as the "tied up," in contrast to wild animals. The herds were the primitive standard of value, as evidenced in the words "pecuniary" (Lat. *pecunia*; from *pecus*, cattle), and "fee" (Anglo-Sax. *feoh*, which means both "gold" and "cattle"). Among the Romans ten sheep equalled an ox. When, in the metal age, copper supplanted cattle as the measure of value, the bars were stamped with images of a cow, sheep, or dog. The waggon, drawn by oxen, was one of the earliest inventions, and we therefore find common names for the wheel and axle. These were made out of a tree trunk cut crossways. The primitive boat was a hand-paddled tree-trunk, which was afterwards scooped-out by stone axes with the help of fire. The common Indo-European word for "tree-trunk" and "boat" is the same, and, in extended meaning, for "sacred tree-trunk" and "temple." For woods and groves—the haunts of man primeval—are the oldest temples in all forest countries. Grimm says, "What we conceive of as a house built and walled-in passes, the further we go into early times, into the idea of holy ground hedged in and surrounded by self-grown trees." Motion is life; where there is life there is spirit; that is primitive man's universal way of looking at things. Trees grow and blossom; bring forth fruit; bleed and moan when cut; wither; and die. An old writer says that "when an oake is falling, before it falles it gives a kind of shriekes or groanes, as if it were the genius, or spirit, of the oake lamenting." Hence the belief that they have souls, a belief shared even by the early Greek philosophers, because of the movements and the sounds which trees make. By stages easy to follow, the tree became a god, whose life was connected with that of the tree. This tree-spirit was next incarnated in a human being, whose life and vigour ruled that of all vegetation, so that were he suffered to live on till he was old and decrepit, the trees and plants would become barren, and the earth would yield no harvest. Therefore he was killed ere that could happen, as among the Congo natives, who, believing that if their chief priest were to die a natural death the world would perish, clubbed or strangled him on the first signs of serious illness, so that his spirit might be transferred to a younger or healthier man. Still more apposite to this was the ancient custom in the grove sacred to Diana, at Nemi, near Rome, the priest of which could secure that office only by killing the man who held it, and had, therefore, to be always on the alert, lest he was slain himself. For he was believed to be the incarnation of the tree-spirit, on whom the fruitfulness of the soil depended. If he became decrepit, then the earth would become unfertile; so, when he was unable to hold his post against attack, he was killed, and his power and vigour, as the tree-spirit, was believed to pass into his successor. Strange as this reads, it is related to a vast body of customs which attended, and still attend, the worship of tree-life, to the bringing of offerings, to the holding of festivals, and to the belief in trees as connected with oracles and divination. The Indians will consult their sorcerers as to the message of the tree which shook the most; the Hebrews had their tree of "soothsayers" and "revealers," and the prophetess Deborah gave her answers under a palm near Bethel which tradition marked as the grave of the nurse of Rachel — one of many holy trees by burial places—while among the Greeks the rustling of the leaves of the sacred oak of Zeus at Dodona was interpreted as the voice of the god. In neolithic times that tree held the supreme place in nature-worship in Europe, doubtless because it formed most of the dense forests, and it was by the rubbing of its wood that the sacred fires were kindled. The ash ranked next to the oak in the honour paid it. But an end must be made of references to a tempting subject which further shows that we cannot handle any side of primitive belief without touching every other side.

Returning to a few more examples of the evidence from language, we find in the identity of the word for "plough" and "branch" that the original implement was a crooked bough. Certain parts of it were named after animals, as in Sanskrit, where the word for "wolf's teeth" and the "plough," as a tearing implement, is the same; and in Old Irish, where *soc* means "ploughshare" and "pig's snout," an appropriate term for a machine used in rooting-up the closely-matted forest soil of Europe. The words for barley, wheat, and flax agree in all the Aryan languages. The names of the camel and the ass are loan-words; that is, they were borrowed from non-Aryan speaking people by the Asiatic Aryans, and so passed on to Europe. For both animals are natives of the East, and their taming, as probably also that of the horse, was effected, by the nomadic races of the wide tracts of Central Asia or the deserts of Arabia. The names of the lion and tiger are absent from the Aryan languages; the larger cave-lion was extinct long before neolithic times, but the range of the living species included south-eastern Europe within historical times. The conclusion from this is that neither lion nor tiger was known to the primitive Aryans.

By the help of the materials of which the foregoing are only random samples, some picture of the culture of the Aryans or Indo-Europeans is possible. As indicated, the long-accepted theory of their origin in Asia is practically abandoned in favour of the theory which places their primitive home in some locality between the Baltic shores and the Ural mountains. It is by no means improbable that their remote ancestors were the men of the Danish shell-mounds who in the course of thousands of years spread themselves over Europe. Miserable was their first condition; as low as that of the Chauci of the German fenlands thus graphically described by Pliny: "In their huts upon the banks they look like sailors aboard ship when the tide is in, and like shipwrecked men at the ebb; and they hunt the fish round their hovels as they try to escape with the tide. They have no cattle, and so they cannot live on milk like their neighbours, nor can they even fight with wild beasts when every stick is carried out to sea. They weave fishing-nets out of sea-tangle and rushes, and they pick up handfuls of mud, which they dry in the wind—for they have not much sunshine, and so they make a fire to scorch their food, and their bodies too all stiffened by the cold of the north." From the loins of these rude, hardy folk sprang some of the makers of the English nation, and as the men of the kitchen-middens had made the first start in domesticating animals by taming the dog, they would by degrees subdue the animals which roamed the thickets, yoke the oxen to their rude waggons, and cut their way through forest with their roughly ground axes. Clearings would be made in the tangled woods; the tree-plough would tear up the soil in which the seeds were to be planted, and, with them, the seeds of a settled condition which, among these Neolithic people, reached its highest stage in the lake-dwellings. There the "tied-up" animals formed their staple wealth in supply of clothing, meat, and milk. For stronger drink they had mead, prepared from wild honey, a liquor still known in rural districts. In Northern Europe this was supplanted by beer, the name being retained in our word "ale" (Old German *alu*, "mead"). It is interesting and amusing to find Pytheas, the "Humboldt of antiquity," in describing British beer, remarking that the Greek physicians warned their patients against it, as "producing pain in the head and injury to the nerves."

The pastoral life, with its care of flocks and herds, and easy provision of food and clothing which these afforded, is far less laborious than the tilling of the soil, with its uncertainties and disappointments. Only the pressure of population and the resulting insufficiency of pastures, which cannot support beyond a given number; compelled the cultivation of the land. Man is by nature and habit a roving animal; as late as the times of Thucydides (fifth century B.C.) the Greeks are described as filled with the nomad instinct; and Strabo, writing four centuries later, speaks of the Germans as careful only to provide for the needs of the day, "to live in waggons and move their cattle where they like."

The social relations were of course more developed as settled conditions were reached, but before any great divergence of the Aryans occurred the several degrees of kinship by blood and marriage were already recognised. The family is the unit of society, not only when the ties between the parents are loose and temporary, so that descent is reckoned on the mother's side, but also when totemism, or scarcity of women, leading to their forcible capture, rules the choice of mates. Among the Aryans wives were obtained first by seizure; afterwards by purchase; a survival of which latter custom is seen in the occasional sale of a wife among the lower orders, under the delusion that the transaction is valid. The husband and father had supreme power over the household; at his will the useless aged were killed; and with him rested the decision whether or not the newborn babes should be reared or "exposed," a custom which survived among Norsemen, Gauls, Greeks, and Romans. It also prevails to this day, more especially in the murder of female infants, among semi-barbarous people, as the Chinese, and among wholly barbaric tribes. The wife was the husband's property, and custom—based upon those supposed needs of the dead which have been already referred to—compelled her self-slaughter at his grave, and also the killing of slaves, who, as prisoners of war—the great nation-maker—were divided among the victors.

The collection of families into "gentes" or bands of "blood-brothers," must have taken place at a very early period. At the bottom of the scale it was never wholly the selfish code: "each for himself, and the devil take the hindmost." Among the men of the river-drift and the cavern the interests of the family made for unity, and subordinated its welfare to that of the gens. The gentes combined into tribes under a *rex* or "ruler," who led them in war and saw that the law—based on custom, the same word originally denoting both—was duly kept. The "moot" or meeting-place was on some natural or artificial hill where, perchance, the dead lay buried; where the altar was reared, and the rude temple raised above it. The connection of law and religion has been intimate from ancient times. Mommsen tells how the representatives of justice assembled on the Mount of Alba when the Latins gathered for religious festival; and in the Brehon Laws (the ancient code of Ireland) the rare penalty of death was inflicted upon the wilful disturber of "any lawfully constituted assembly." The labour of clearing the ground for cultivation was shared by the "village community," aided by slaves, and therefore at the outset the land was public property, the establishment of private rights based on a man's "realty" (or "reality") being later.

To enlarge this sketch would be to attempt the impossible task of compressing the history of civilisation among settled folk into a few pages. For although the Indo-Europeans represent only one among other civilising peoples, and are, in fact, largely indebted to these, their history is that of the nations who have most profoundly influenced, or whose successors most profoundly influence, the destinies of mankind. It is this fact that has warranted the prominence given to them in this book of outlines, while the highest interest attaches to them as possibly later links in an unbroken chain of man's occupancy of Europe.

Brevity is apt to mislead, so let the caution be given that the foregoing summary of the leading features of Aryan civilisation does not imply a uniform stage among the several peoples. Very complex and subtle are the causes arresting or quickening progress, causes lying too remote for analysis; but we know from the relative state of civilisation of Celt and Teuton, and of Greek and Roman, at the dawn of history, what advances the southern peoples had made, as compared with the northern.

These latter, as we learn from the narratives of classic writers, and from the known course of events in north-western Europe, had not passed beyond the semi-barbaric stage centuries after the Greeks had defeated the Persians under

Xerxes at Salamis, and after the Romans had laid the foundations of their mighty empire on the Palatine Hill. Very early, therefore, as the need of more elbow-room drove the various branches further afield, the ancestors of the Mediterranean peoples had migrated from the northern region described in the Odyssey, among other memories of a barbaric past, as the land where the Cimmerians dwell "shrouded in mist and cloud, and never does the shining sun look down on them with his rays, neither when he climbs up the starry heavens, nor when again he turns earthward from the firmament, but deadly night is outspread over miserable mortals."

Excavations carried on during the last few years are revolutionising old theories as to the origin and antiquity of European civilisation, showing that this ran on lines of independent development, although not isolated from the great civilisations of the East. Compared with what we now know, the Homeric age is modern. Dr. Arthur Evans's discoveries in the island of Crete bring to light a centre of culture older than that which Dr. Schliemann revealed at Tiryns and Mycene, which perished somewhere near the middle of the twelfth century B.C., through the onslaught of Dorian invaders who sacked and gave both cities to the flames, the survivors of the catastrophe finding new homes in Ionia. Crete, a stepping-stone from Greece to Asia Minor, and centre of communication between Egypt and the western Mediterranean, was, Dr. Evans affirms, "the true cradle of European civilisation." Tradition centres round the "sea power of Minos," hence, not inappropriately, the term "Minoan" is applied to a stage of culture which appears to be in unbroken connection with the Neolithic Age in Europe, carrying us back millenniums before the Iliad and the Odyssey were written. Much obscurity hangs round the questions of race and language, because the Cretan inscriptions are without any such bilingual key as, in the case of the famous Rosetta stone, enabled scholars to decipher the Egyptian hieroglyphics.

These discoveries in the Ægean relegate the activity of the Phœnicians, as the distributors of wares between East and West, to a secondary place; the enterprise of Crete being manifest in use of a commercial script along the Mediterranean 5,000 B.C. For Dr. Evans's explorations establish the fact of active intercourse between Crete and Greece, Egypt, Syria, and other countries, centuries before the Phœnicians launched their craft upon the midland sea and trafficked with Cypriote and Cretan, or sailed beyond the Pillars of Hercules for the coveted tin of Britain. (See Chapter IX. *Story of the Alphabet.*)

The origin of man in America is an insoluble problem, but much is to be said in support of the theory that he reached there both by way of Behring Straits and by creeping along the coast of Indo-Malaysia and Eastern Asia. The Indians still voyage hundreds of miles in frail canoes, and what such folk can do is the measure of capacity of barbaric man in the past to have done likewise.

Such migration from the shores of the Indian Ocean, if it took place, reaches back to a time when Palæolithic man was living in Europe, and when Asia was still in the Stone Age, because, among other reasons, a lapse of time long enough for the development of the ninety languages spoken by the North and South American-Indians is required. As for Africa, Egypt, (which seems scarcely to belong to her), and the northern sea-board excepted, the races there, isolated by the desert wastes, must during pre-historic times have worked out their slow but soon arrested advance above the lowest savagery.

CHAPTER V.

THE AGE OF METALS.

FOLLOWING the art of fire, the two chief agents in man's progress have been the discovery of metals and of the power of steam. Upon this later discovery of the two, only the passing remark can here be made that until the power of steam was known the means of transit had not sensibly improved since pre-historic times.

The revolution wrought by metals is the greatest that the world has yet seen or that it will ever see. No wonder that round these treasures of the earth, gathered from the glittering sands or melted from the deep rock, there gathered many a myth and legend which ascribed them to the gods; which told of the wondrous smiths—as Tubal Cain, Hephæstus, or Vulcan, Ilmarinen, and Wieland or Wayland—and which filled the underworld with cunning workmen, gnómes and fairies, and guardians of hidden riches.

It is said that the origin of paying tithes on minerals is due to the belief that the ore is a living organism. The clergy of past times either shared or fostered the delusion, because tithe was only payable on products of the earth which renewed themselves annually, and these were said to include ore in the vein.

In Finnish myth iron was said to originate from sprouts, probably through the observation that bog-iron ore, which has a spongy texture, assumes tree-like forms. "When once this was assimilated in thought to the vegetable kingdom, it had to be watered and nourished like any other plant. This gave rise to the development of the story by an incident in which the daughters of Nature spilt milk upon a marsh."

When the neolithic people drove their piles into the lakes of Switzerland and Italy and Thrace, Europe was covered with dense forests. The oak and the fir grew as far south as Greece, the vine had long flourished in Southern Europe, but the olive had not yet been brought from its Syrian home. Only when and wherever there was placed in man's hand the hard, sharp-edged bronze or iron axe could he make quick clearance of the trees and hew his path to that goal of civilisation which he could never have reached by stone implements.

Controversy has gathered about the unimportant question whether gold or copper was the earliest metal known. River gold would attract attention, and it may have been known to the primitive Aryans; but the more frequent outcrop of copper in certain regions, where traces of early settlements occur, point to it as the primitive metal. Its use in a native state everywhere preceded its use as a compound in the form of bronze. Although generally found in small quantities, large masses have been obtained from Russia and from the shores of Lake Superior, where blocks weighing several hundred tons have been more than once discovered. As shown already, North America supplies the best evidence of a Copper Age, the mound-builders breaking off the metal as needed and hammering it cold into the required shape, modelling this on the stone tools and weapons used side by side with it. "Accustomed to the use of stone, they would at first regard the metal as merely a stone of peculiarly heavy nature, and on attempting to chip it or work it into shape would at once discover that it yielded to a blow instead of breaking; that in fact it was a malleable stone. Of this ductile property the North American savage availed himself largely, and was able to produce spear-heads with sockets adapted for the reception of their shafts by merely hammering out the base of the spear-head and turning it over to form the socket, in the same manner as is so often employed in the making of iron tools." That he should not have smelted the metal is curious, because the fires upon the altars were intense enough to melt down the copper implements and ornaments laid upon them, so that the mound-builders "may in one sense be said to have been in an age of stone, since they used the copper, not as metal, but as stone."

Turning to the old world, copper celts of a

simple form have been found in Central India, Switzerland, and Hungary, and, as should be expected, since that island has given its name to the metal, in Cyprus. One specimen found in an Etruscan tomb, and now in the Berlin Museum, is of the shape of an ordinary stone celt, having been probably cast in a mould formed upon the more primitive implement. That the new is the offspring of the old is well exampled in the story of the archæologist who picked up a fine bronze celt and carried it off in high glee to a brother antiquary. The friend looked at it, and said, "Yes, it's a good specimen, and old too; but I've something older: the mould in which it was cast," and bringing out his find, fitted the one into the other.

But the use of copper in a native state was only intermediate. Its real utility began with the discovery that by mixing a certain proportion of tin (or in some instances zinc, lead, or, more rarely, silver) with it, a hard and tough material resulted. Where and when this simple but important discovery was made is unknown. Common opinion leans to the theory that bronze was introduced into Central Europe about 1500 B.C. from the Mediterranean through Phœnician traders. But if this be so, we are no nearer knowledge of the source whence Phœnicia obtained her copper supply. It appears not unlikely that bronze may have reached Europe by a more northerly, or north-easterly route, as abundant remains of copper procured from the Ural and Altai mountains have been found in the "kurgans" or burial mounds of the semi-legendary "Tschudes," scattered over the plains of Northern Asia. As regards tin, there is well-nigh common agreement that the Cassiterides or "Tin Islands" (Greek *kassiteros*, tin), which are identified with Cornwall and the Scilly Isles, were an early and abundant source of supply to the Phœnician voyagers. An interesting discovery was made some few years ago in the dredging up of a large H-shaped ingot of tin in Falmouth harbour. Its form adapts it for being laid in the keel of a boat or slung on a horse's side, two ingots thus forming a load for a pack-horse. From certain references in the old geographers it seems that the tin was carried from Cornwall to the Isle of Thanet, which was then part of the mainland, and thence shipped by short passage to Gaul. "This," as Professor Rhys remarks, "would explain Cæsar's singular statement that British tin came from the inland parts of the country."

Three methods of casting bronze implements were adopted. The earliest was to pour the molten metal into a pattern cut into stone or into an impression made by a stone celt on a mould of loam or clay. An advance upon this would be a double mould of these or some harder material; and the next step would be to that of a solid mould by placing a model of the implement in wax, wood, or other combustible substance, and encasing it in some porous non-combustible substance. The exposure to fire would burn out the model and leave a cavity into which the molten metal was poured, reproducing the design. Although the reference lies outside our subject, visitors to Pompeii may remember that by pouring-in liquid plaster of Paris into the hollows occupied by the skeletons of the victims of the eruption, and letting it harden, perfect casts of the bodies in the clothes which they wore when overtaken by the engulfing ashes were obtained. The bronze workers ornamented their decorated objects with circles, spirals, and zigzags of a more or less uniform design, figures of animals or plants being rarely attempted. But the skill and symmetry shown in the treatment of the metal evidence no mean advance in art. The fragile moulds in burnt clay are seldom found, but more or less perfect single moulds are not uncommon.

The hatchet or axe, known as the celt, is the most abundant type, and was used both as a tool and a weapon. There is the flat or simple kind; the flanged, or ribbed kind; the winged, or with flanges extended so as to form a socket for the handle, and the socketed kind, and of each kind there are several varieties. The flat celts are the earliest, approaching in character, as already observed, to the stone hatchet, and both these and the more elaborate kinds sent antiquaries of the old school on the wrong tack, just as the stone celts and arrow-heads puzzled both ancients and moderns.

Speaking of discoveries near St. Michael's Mount, Camden, who lived in the sixteenth century, says: "At the foote of this mountaine within the memorie of our Fathers, whiles men were digging up of tin, they found Spearheads, axes, and swordes of brasse [bronze, by the way, is always translated "brass" in the Old Testament], wrapped in linnen, such as were sometimes found within the forrest of Hercinia, in Germanie, and not long since in our Wales. For evident it is by the monuments of ancient Writers that the Greeks, the Cimbrians, and the Britans, used brazen weapons, although the

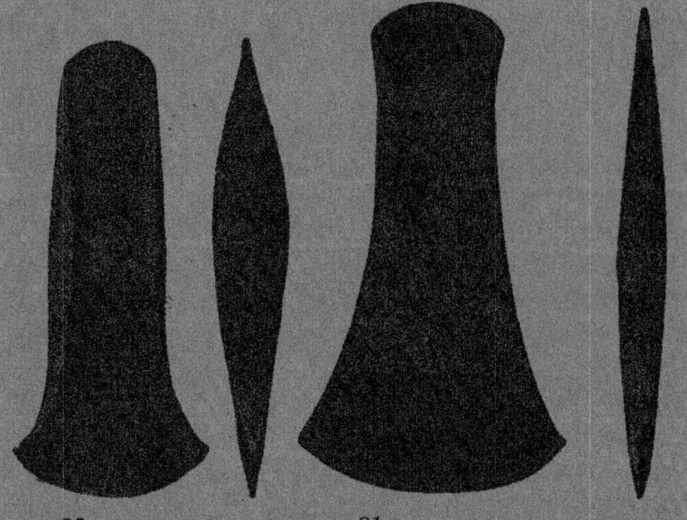

FIG. 80.—FLANGED CELT, NORFOLK. (*Evans.*) FIG. 81.—FLAT CELT, BUTTERWICK, YORKSHIRE. (*Evans.*)

wounds given with brasse bee lesse hurtful, as in which mettall there is a medicinable vertue to heale, according as *Macrobius* reporteth out of Aristotle. But happily that age was not so cunning in divising means to mischiefe and murthers as ours is." Others conjectured that they were chisels for cutting stone for the Roman roads; or rests to support the *lituus* or crooked staff of Roman augurs; or tools for the engraving of letters and inscriptions; or the sickles with which the Druids cut the sacred mistletoe. Reference to the fantastic notions of which these are specimens is useful if only as reminding us how far we have travelled nowadays from that sort of thing. From what a different standpoint we look upon the past! The old school had learned its lessons within the four corners of "sacred and profane" history, as it was oddly called, and explained all that was curious and unfamiliar within those limits. But these dividing lines are now effaced, the antiquity and the slow ascent of man demand a margin which melts into the horizon beyond which lie the beginnings.

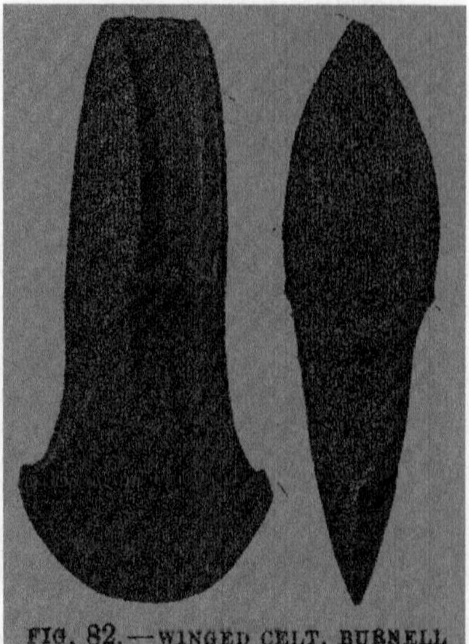

FIG. 82.—WINGED CELT, BURNELL FEN, CAMBRIDGE. (*Evans.*)

It would neither lighten nor brighten this rapid record of the transition from stone to metal if a comparison of the various types of tools, weapons, and ornaments in the several bronze-using countries was added. The new departure, in placing a new material in the hand of the craftsman, developed his ingenuity and fostered that tendency to more artistic finish which presupposes more leisure for the work and more refinement in the worker. The types vary in different countries, those in Britain, for example, showing a certain degree of originality, but the variations can interest only experts. The list of objects figured in monographs on the subject is miscellaneous, comprising, besides the three classes of celts—flat, flanged, and socketed—gouges, tongs, knives, awls, sickles, rapiers, daggers, halberds, spear-heads, buttons, rings, earrings, bracelets, pins, and so forth. Some of the sickles—an implement rare in Britain, perhaps owing to less cultivation of grain there—are elaborately hafted; the swords have pommels inlaid with ivory and amber; and the shields have bosses of bold and rich design. But of some of these a few specimen illustrations are worth pages of description.

FIG. 83.—SOCKETED CELT, REACH FEN, CAMBRIDGE. (*Evans.*)

The value of the relics in the Swiss lake-dwellings as contributions to our knowledge of the metal-using period is especially evident. The villages assigned to the Bronze Age decrease in number as that metal becomes superseded by iron. They are situated chiefly in Central and Western Switzerland, and they differ little in general character from those of the Stone Age, except in being, in some instances, farther from the land, perhaps because construction became easier, and also, perhaps, because with universal improvement in weapons the risks of attack were greater.

The occurrence of highly ornamented bracelets among the relics in the lake-bottoms has led to the suggestion that they may have been thrown into the water as offerings to the spirits whom

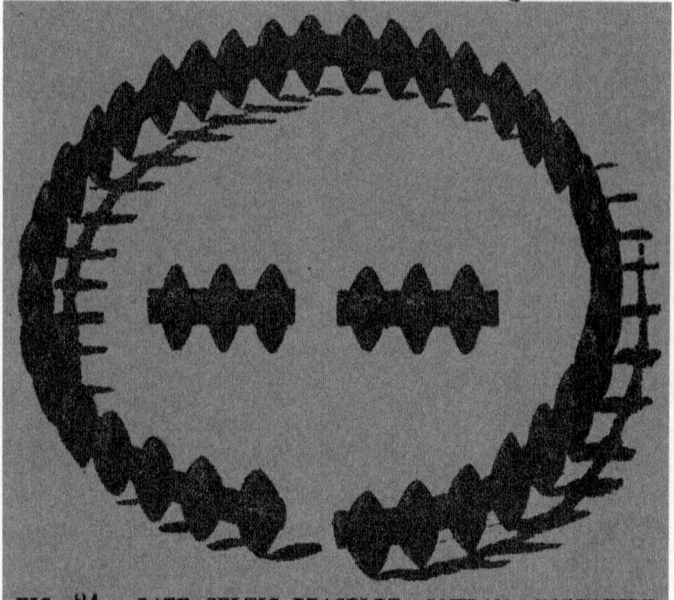

FIG. 84.—LATE CELTIC BRACELET, COWLAM, YORKSHIRE. (*Evans.*)

barbaric people believe dwell in river, lake, and ocean. We have nothing to guide us to definite knowledge of the religion of the lake folk, but there is good reason for assuming that they practised the nature-worship already described, and therefore that they worshipped water as the home both of maliceful and helpful deities.

If the dead, dull stones were believed to have spirits within them, still more so would this seem true of water, whose every motion denotes the presence of life, and of perennial youth and

FIG. 85.—BRACELET WITH CUP-SHAPED ENDS, CO. CAVAN. (*Evans.*)

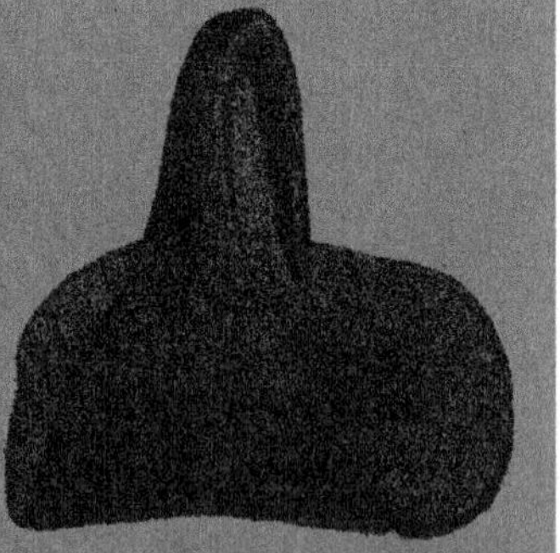

FIG. 86.—BRONZE EAR-RING, COWLAM, YORKSHIRE. (*Evans.*)

strength. The fury of the wind-tossed sea; the boiling and seething of the tumbling cataract; the swirling of the rapid; the swift current of the river; each with its own voice, were all out-

FIG. 87.—BROOCH, COWLAM, YORKSHIRE. (*Evans.*)

ward and visible signs of the spirits within. Helpful or harmful, as the things which they did showed them to be, it was they who seized the drowning as their prey, and swallowed whatever was offered them; who dried up the springs and streams, or who laved the land so that the earth brought forth her increase; who gave food to man in the supply of fish, and who swept away the disease demons and carried

FIG. 88.—SEA GHOST, FROM A NATIVE MELANESIAN DRAWING. (*Codrington.*)

them to the great sea. So the Peruvians believed, when, on the arrival of the rainy and unhealthy season, they drove the fever-bringing spirits with shouts and brandishing of lances to the river-banks, that they might be banished to the ocean. So among the ancient Babylonians, the patients bathing in the Euphrates besought the river to carry away their disease, and "bear it down the stream." So, within the last few years (as perhaps they still do in secret to this day), the Manx peasants repaired to sea and wells for ailments, believing in the miraculous power of the water if used "while the books are open at church," that is, during divine service. Everywhere we meet with a belief in great gods of ocean, sea, and mighty river, and in godlings of the minor streams, pools, and springs.

"A blind people," wrote old Gildas, "paid divine honour to the mountains, wells, and streams," and in the old Celtic religion the rivers were specially identified with certain gods and goddesses. The full rivers were "mothers" who brought them food; in the whirling eddy the water-demons lurked; the lake was ruled by a lonely queen, and every well had its nymph. Among the sea-gods were, among others, Nodens, to whom a temple was built, as late as the time of the Roman occupation, at Lydney, on the banks of the Severn; and Ludd, whose name perhaps survives in Lud-gate; St. Paul's Cathedral, as some think, marking the site of a temple to him. These correspond to Neptune among the Romans, and to Poseidon, Nereus, Proteus, and other gods among the Greeks, who had their nymphs in ordered rank: the Oceanides, daughters of Oceanus; the Nereides, dwellers in cool cave and grotto, daughters of Nereus; and the fresh-water Naiades, nymphs of springs and streams. Thus, so it would seem, did the old nature-worship of the lake-dwellers survive in this and other forms, among Celt and Greek and Roman. For in religions there are no inventions, only survivals. "Father Thames" and "Father Tiber" show how old myths are preserved till they ultimately become figures of speech, and as for the army of water spirits, their names remain as legion: from malicious kelpies to bewitching mermaids. We need not travel to Africa or New Grenada or Peru for examples of sacrifices to the water to ensure a good catch; since we find these among the fisher folk of the Scottish isles, of which a traveller records example when the inhabitants of Lewis, addressing a sea-god called Shony, throw a cup of ale into the sea "hoping for plenty of sea-ware." And the belief that it is unlucky to save a drowning man, thus cheating the water-demon of his lawful prey, is not confined to the Hindus and Malays and Kamschadals, but is found among Bohemians and St. Kilda islanders. In Scott's "Pirate," Bryce the pedlar warns Mordaunt against saving a shipwrecked sailor: "Are you mad," said he, "you that have lived sae lang in Zetland, to risk the saving of a drowning man? Wot ye not, if you bring him to life again, he will be sure to do you some capital injury?"

As illustrating the identification of rivers with certain divine beings among the Celts, there are those that still bear the name of Dee or Deva and kindred ones. "As late as the time of Giraldus the Dee was supposed to indicate beforehand the event of the frequent wars between the Welsh and English by eating away its bank on the Welsh or on the English side, as the case may be. The name of another river marks it out as one that was formerly considered divine, the Belisama, probably our Ribble."

But it is more especially round wells that superstition has gathered and lingers to this day. The sacred rivers of the world, cleansing the soul of sin and bearing it away like some foul disease, are many: the Ganges, best-known type of all; but the healing spring, often in virtue of its curative properties, and perhaps, as connected with the rite of baptism, has been the most fertile source of long-enduring beliefs. The wise and tender treatment of the old nature-worship by the Church is notably manifest here. The name of madonna or saint, or the general term "holy," was substituted for that of the godling or spirit of the spring, and the several rites connected with water-worship maintained under thin disguises. Barbaric lustrations reappear in the sacrament of baptism; the brush once used in the pagan temple is dipped in "holy water," and sprinkles the faithful. The leprous Naaman repairing to the Jordan; the sick waiting their turn to be dipped in Bethesda when the spirit "troubled" the water; have their modern representatives in the halt, blind, dumb, and otherwise afflicted, who repair to St. Winifred's Well in pathetic hope of cure by bathing in or drinking of the stream. Legend tells how this burst forth from the spot where the head of the martyred saint fell, as did the triple fountain outside Rome when the head of the apostle Paul struck the earth three times. Round the altar of the chapel which encloses the well the sufferers kneel in prayer for St. Winifred's aid, and, when the festivals in her honour are held, kiss with fervour the reliquary that holds her bones. The crutches of the cured hang on the chapel pillars like votive offerings in classic temples, and link the worshippers of to-day with the neolithic villagers who cast their gifts, as old writers tell, into the lakes of southern France and central Switzerland.

Space allows only brief allusion to the symbolical use of water, as in "wishing-wells," "cursing-wells," and so forth. This is a department of that widespread activity of barbaric thought in what is known as "sympathetic magic," or the supposed vital connection between object and subject. Hence to this very day the betrayed girl or the injured man will make a rude clay image of the wrongdoer and lay it in the stream, so that, as it washes away under the play of the running water, his life may waste to its doom. Or his name will be scratched on a pebble and curses uttered against him as the stone is thrown into the water; or pins will be cast in, so that, like as with a dagger, his heart may be pierced. Sacred trees are closely associated with wells, and on those growing near them the sick hang rags or shreds of clothing as vehicles for the transfer of their diseases. "By the intercession of the Lord, I leave my portion of illness on this place," is an old formula used by the Munster peasants. Then there is the great body of portents and omens drawn from the appearance of the water, as the reading of health or sickness in the ripples; the counting of the bubbles that break when anything is thrown into a mineral spring, the lovelorn reckoning thereby what months or years must pass before the absent one returns, or the wedding takes place. Of ghastly ordeals, too, water has been the vehicle in the throwing of suspected witches into it. If the poor creature sank, there was evidence that Satan had claimed his own; sometimes by cruel irony, if she swam, there was evidence that the sacred stream rejected the criminal. And there is the pathetic idea of the relation between water and death. Life, as many a rustic yet believes, goes out on the ebb, a superstition of which Dickens makes use in the death scene in "David Copperfield": when "it being low water, he went out with the tide."

The subject, charged as it is with the primitive ideas of man concerning his relation to everything around him, tempts us to linger over it, and perhaps such examples as have been chosen out of many will help us to realise the continuity of belief and similarity of custom which have prevailed since the dawn of thought, and how the old persists throughout the new, because the roots of each are in the same soil. One of the most striking illustrations of this is in the action of an ecclesiastical council held at Durham in 1220, when it was ordered that the fonts were to be kept locked because people were in the habit of carrying away the baptismal water for use in magical rites. Many of our mediæval fonts show where the hinges and staples for the lock were formerly fastened.

Besides the bracelets which have led to these remarks, there have been found carved and curiously shaped stones which, in the judgment of some writers, point to other phases of nature-worship. But to dwell on this would be only to amplify examples of the varied outward forms of one underlying idea. To borrow a figure from chemistry, we may call religions allotropic. That is to say, as the diamond and a piece of charcoal are different combinations of the atoms of the same substance, so religions are the same in essence, and vary only in the mixture.

The wellnigh common practice of burning the dead in the Bronze Age was probably resorted to as a yet more effective way of getting rid of the ghost than by burial of the body, with the placing of heavy weights upon it, and a ring fence around it. Cremation seemed akin to annihilation, and as the connection between corpse and ghost was a vital one in barbaric belief, the bodiless, homeless spirit was less likely to be troublesome. While the primary object of the custom was doubtless to completely destroy the connection between body and soul, it would be especially adopted by nomadic peoples, who, in leaving their dead behind, would be unable to make provision for the appeasing offerings at their graves. Hence the burning of the body to prevent the neglected ghost from following and harrassing the living.

The old conception of the tomb as the home of the dead, however, remained. A number of so-called "house-urns" have been discovered in various parts of Europe which, with differences in detail, resemble each other as receptacles for the ashes gathered from the funeral pyre. They are imitations of the beehive, oven, or roundish huts of primitive Indo-European type, "the walls of which we must imagine to be composed of loam, twigs, or other perishable material. The roof seems to have consisted of layers of straw or reeds, and to have been held together by ribs, which in the real house were made of wood." In some of the German funereal urns the door is shown in the roof, having been put there in the original dwellings as a protection against wild beasts; the urns with the door in the wall representing a later style of house. One urn is obviously a copy of a lake-dwelling, with added ornamentation in the double spirals which are a characteristic pattern of the Bronze Age. Wholesome and sanitary as is cremation, it has the disadvantage of preventing the preservation of articles throwing light on the dress and habits of a people, but, fortunately, a few interments supplement what has been gathered already on these matters from the lake-dwellings of the Bronze Age.

In his account of the succession of the metals, Schrader refers to the struggle in early times between gold and silver for supremacy. The "white" metal, as it is named in many languages, both Aryan and non-Aryan, was for a long period preferred to gold, the "yellow" metal (root *ĝhel*, yellow), probably as less widespread, being found only in rocks, and hence harder to extract. The Turko-Tataric name for silver comes from a root meaning "to hide," indicating that it was difficult to get at. The trade routes by which it was conveyed from place to place are uncertain, but probably the rich mines of Armenia were a leading centre of distribution both east and west. Speaking of the German tribes of the interior, Tacitus says that "they prefer silver to gold, not from any special liking, but because a large number of silver pieces is more convenient for use among dealers in cheap and common articles." That lead was known before iron is shown by the results of Dr. Schliemann's excavations in deposits of the Bronze Age at Mycenæ, where a good many discoveries of it were made. And that copper preceded iron is evidenced, among other reasons, in the application of its name to the later metal. For example, the German *eisen*, the Celtic *eisarn*, and the English *iron*, are all related to the Latin *æs*, copper. The ancient Mexicans also used the same word for both metals. Some of the names given to the smith's tools show that they were originally of stone. Hammer is the Anglo-Saxon *hamor*, a rock, and the Greek words for "smith" and "smithy," are derived from the word for copper. If *sidēros*, the Greek word for iron, is connected with *aster*, a star (*cf.* Latin *sidus*, a star), this may show, as we know to be the case among the Esquimaux and some other barbaric races, that that metal was first known in meteoric form, as composing a large proportion of the bodies falling from space, named aërolites, or "air-stones." The Coptic name for iron is said to mean the "stone of heaven," and a like idea is found in the Finnish epic, the Kalevala, where red milk.

flowing from the breasts of the Rainbow maidens, becomes Iron, who goes to see his brother Fire. Fire tries to devour him, but he escapes to the swamps, where the divine smith Ilmarinen captures him and makes him into "swords of heroes and buckles for women." Iron was smelted on the shores of the Black Sea and in other parts of the East in the time of Homer (about 850 B.C.); how much earlier than this the metal was used in Asia is unknown. As already stated it is spoken of in the Iliad and in the "Works and Days" of Hesiod, a poem nearly contemporary with the Homeric, as of rarity and value. By what channels the use of it spread through Europe is not clear, but its diffusion must have been rapid, because when the Romans were attacked at the beginning of the fourth century by the Gauls under Brennus, they found these despised barbarians armed with iron swords. And although it is said that some of the Saxons who fought under Harold at the battle of Senlac in 1066 wielded stone malls or hammers, this was probably due to scarcity of metal weapons. Be this as it may, with the appearance of iron tools and weapons the story of primitive man is ended.

In the "Age of Iron" there is no place for fairies and witches. Where pieces of that metal are hung, they dare not venture: hence the horse-shoes still to be seen nailed on stable doors and masts of ships, although they are now considered as luck-bringers. Aubrey tells us that "in the Bermudas they use to put an Iron into the fire when a Witch comes in," and he adds,

"Under the Porch of Staninfield Church, in Suffolk, I saw a Tile with a Horse-shoe upon it, placed there to keep the Witches away, though one would imagine that Holy Water alone would have been sufficient."

CHAPTER VI.

CONCLUSION.

THE foregoing outline of man's early history shows that his path throughout is strewn with the *rude tools and weapons* wherewith he carved and fought his way. And it also shows that civilisation retains, and, in no small degree, shares *his primitive* ideas about his surroundings. Thus the one tell us what he *did*, and the other what he *thought*. And as, under the same conditions, he made shift with the same stone implements, so, at the same level of culture, he explained like things in a similar way, and behaved towards them accordingly. To the confirmation of this the habits and beliefs of barbaric races of to-day contribute much.

In these examples, both from past and present, of like use of things and of like modes of thought, the unity of mankind is further brought home to us. We have not altered so much as we vainly think; if the civilised part of us is recent, in structure and inherited tendencies we are each of us hundreds of thousands of years old. And the explanation of this is that the influences of a few generations, acting on us from without, are slight contrasted with the influences of a dateless past inherited from our ancestors. These explain our mental as well as our bodily vestigial structures.

But there are other facts which are not so easily brought home to us. 1. There are the vast ages which separate man at his lowest, say as a chipper of flints in the Thames and Somme valleys, from Plato and Shakspere, from Aristotle and Newton, or from the highest types living. Yet through all those ages, man, as we now know him, was in the process of making; he is their result. 2. There are the long, monotonous ages, the dismal wildernesses strewn with his bones, before any great impulse to advance, such, for example, as came through knowledge of metals, was given. Till then, brain and hand carried man only a certain distance, where progress was arrested. Nevertheless, all the germs were there, and we can trace their development. From unhafted stone tools and weapons to modern machinery; from clothes of twisted grass to modern fabrics; from tattooing and shell ornaments to adorning with rare jewels; from wigwams and mud-huts to mansion and palace; from heap of burial stones to pyramid; from cromlech to Buddhist tope and Christian cathedral; from scratchings on bones to sculptures of Phidias and paintings of Raphael; from twanged bow-string to violin of Stradivarius; from picture-writing to alphabets; from signs and imitative sounds to rich vocabularies; from counting on the fingers to reckoning in trillions; from measuring with different parts of the body—foot, nail, hand, span, fathom—to geometry; from "dug-out" canoe to Atlantic liner; from calabash or horn and clay-smeared vessel to Sèvres ware; from family life to tribal and national unity; from work done by one to division of labour—the ruler, the farmer, the merchant, the soldier—from coarse rites and bloody sacrifices to ancestors, local gods and nature-gods, to worship of a Supreme being; from myths and such-like guesses to science and its certainties; as from alchemy to chemistry; and from astrology to astronomy; these and aught else in numberless gradations, are the steps of advance by which the highest races have made all things their ministering servants.

3. But there is the fact that just as the lower organisms remain as they were from the beginning, and have all along constituted the majority; so the races which have reached a certain stage and have never passed beyond it far outnumber those above them. For progress is a modern idea —a Western idea. The Orientals, from whom we may now except the Japanese, hate it. As Sir Henry Maine says in his "Popular Government," "the entire Mohammedan world detests it. The multitude of coloured men who swarm in the continent of Africa detest it. The millions on millions of men who fill the Chinese Empire loathe it, and, what is more, despise it." And as Walter Bagehot says in his brilliant little book, "Physics and Politics," "any progress is extremely rare. As a rule, a stationary state is by far the most frequent condition of man, as far as history describes that condition; the progressive

state is only a rare and an occasional exception," Barbaric races, and, for that matter, civilised races, only in lesser degree, are bound hand and foot by custom, by taboo, by belief in luck and ill-luck paralysing action; and are deterred from doing this or that because of supposed divine sanctions or prohibitions, which thus create a hedge round abuses, illusions, and frauds. Moreover, all men are mimics. Captain Palmer relates of the Fiji that "a chief was one day going over a mountain path followed by a long string of his people, when he happened to stumble and fall; all the rest of the people immediately did the same, except one man, who was set upon by the rest to know whether he considered himself better than his chief!" And it is only a few years ago, among ourselves, when a royal personage had an affection of the knee which caused her to walk lame, that "society" affected what was nicknamed the "Alexandra limp." That Fijian who refused to fall because his chief had stumbled represented the rare souls who have dared to suggest that such and such a thing may have two sides to it, and who have thus opened the way to inquiry wherein alone is the path of progress. Their fate has been the crucifix and the stake, or long, pining years in dungeons. Both vested interests and apathy have been the foes to advancement, so strong is the reluctance to change, so great the "pain of a new idea"; so dominant the power of feeling over reason, of that wish to believe which demands no effort against that desire to know which involves strenuous inquiry and application. Yes; progress is a modern idea. We have only to think of the state of France before the Revolution, and of the barbarous condition of the Highlanders—a century ago; of the state of England—three-quarters of a century ago—before social reforms were set afoot; of the infamous penal laws in the time of the Georges; of the political disabilities, of the condition of miners and peasants down to our own time; we have only to think of the state of Russia to-day; to learn that progress has only just begun in the wider and nobler meaning that we are attaching to the word. For the increase of knowledge is only an agent of advance in the degree that it sets people thinking about social questions, about the abolition of class privileges; about equality of chance in that struggle for life which becomes ever more acute. So that progress, although too often reckoned in terms of expansion of trade and wealth, of inventions of new machinery, and of discoveries of new resources of the earth, is, happily, coming to mean more than these, even the disappearance of the artificial state which makes life hideous and hopeless to the many, and the creation of a public sentiment which shall hold the worker in honour, and regard only the wilfully idle as disgraced. Nor this alone. The more equitable distribution of the means of life is an object to be further striven after in the degree that it contributes to that well-ordered use of opportunity in which education is the chief directing power. It is true that the history of both individuals and nations, while it in large degree explains the causes of growth, of arrest, and of decline, also confirms the lesson of man's impotence before the great forces of nature. "Limits we did not set condition all we do." But within those limits there is such room for the play of human activity that the wise expenditure of this should be our chief solicitude.

"No eye could be too sound
To observe a world so vast,
No patience too profound
To sort what's here amass'd;
How man may here best live no care too great to explore."

SELECTED LIST OF BOOKS.

CHAPTER I.

Darwin's *Descent of Man.* Murray, 2s. 6d.
Huxley's *Man's Place in Nature, etc.* Macmillan, 4s.
Clodd's *Story of Creation.* Longmans, 3s. 6d.
(In *The Nineteenth Century*, April, 1895, Professor Prestwich summarises the discoveries of rudely worked flints on the high chalk plateaus of Kent, by B. Harrison. These point to the existence of pre-glacial man in Southern Britain, and, therefore, as removed from Palæolithic man by the long interval "needed for the advance and retreat of the great ice stream.")

CHAPTER II.

Bonney's *Story of our Planet.* Cassell, 31s. 6d.
Geikie's *Class Book of Geology.* Macmillan, 4s. 6d
„ *Text Book of Geology.* „ 28s.

CHAPTERS III. & IV. SECTIONS 1–3.

Allen Brown's *Palæolithic Man in North-West Middlesex.* Macmillan, 7s. 6d.
Evans's *Ancient Stone Implements.* Longmans.
Dawkins's *Cave Hunting.* Macmillan.
„ *Early Man in Britain.* Macmillan.
(Both works out of print.)

Joly's *Man before Metals.* Kegan Paul, 5s.
Lubbock's *Pre-Historic Times.* Williams & Norgate, 18s.
Stevens's *Flint Chips.* Bell and Sons. (Out of print.)
Worsaae's *Pre-Historic Antiquities of the North.* Kegan Paul, 7s. 6d.
Worthington Smith's *Man the Primeval Savage.* Stanford, 10s. 6d.

CHAPTER IV. Sections 4–8.

Codrington's *Melanesians.* Clarendon Press, 16s.
Elton's *Origins of English History.* Quaritch, 20s.
Greenwell's *British Barrows.* Clarendon Press, 25s.
Keller's *Lake Dwellings.* Longmans. (Out of print.)
Nadaillac's *Pre-Historic America.* Murray, 16s.
Payne's *History of America*, vols i. and ii. Clarendon Press, 12s. 6d. each.
Rhys's *Celtic Britain.* S.P.C.K., 2s. 6d.
Schrader's *Pre-Historic Antiquities.* Griffin, 21s.
Spencer's *Principles of Sociology*, vol. i. Williams and Norgate, 21s.
Taylor's *Origin of the Aryans.* Scott, 3s. 6d.
Frazer's *Lectures on the Early History of the Kingship.* Macmillan, 10s. 6d.
Tylor's *Anthropology.* Macmillan, 7s. 6d.
" *Primitive Culture*, 2 vols. Murray, 21s.
Clodd's *Myths and Dreams.* Chatto, 3s. 6d.
" *Story of the Alphabet.* Hodder & Stoughton, 1s.
" *Animism: The Seed of Religion.* Constable & Co., 1s.

GENERAL INDEX.

UNWIN BROTHERS, LIMITED, PRINTERS, WOKING AND LONDON.

Printed in Dunstable, United Kingdom

91219493R00121